I0875883

Affordability 2.0

In a world where complex challenges like healthcare, housing, education, and government affordability dominate the conversation, traditional approaches often fall short. This book introduces a transformative framework for thinking and problem-solving, designed to address the pressing issues of affordability in modern society. By integrating value, customer needs, and cost, this approach moves beyond conventional methods to deliver sustainable solutions that balance performance, profitability, and prosperity.

Affordability 2.0: A How-To Roadmap for Application and Implementation provides a comprehensive roadmap for leaders and managers seeking to transform their organizations and solve complex problems and provide products and services that are affordable for targeted customers. The book covers essential topics such as leadership, strategy, systems thinking, design for affordability, implementation, and sustainment. It features real-world examples, including the F/A-18 Super Hornet, MRAP military vehicles, NASA's Artemis II, Boston Whaler, and numerous other case studies to illustrate the practical application of the framework. Written by the originator of the Affordability concept, this book incorporates decades of experience, proven tools, and techniques to help organizations achieve measurable improvements in performance and cost efficiency. It was written to address the growing demand for innovative solutions to affordability challenges across industries and society.

This book is ideal for leaders, managers, and decision-makers in industries such as healthcare, government, education, defense, and manufacturing, as well as anyone interested in solving complex problems in their organization or community. Whether you are looking to improve processes, increase profitability, or drive sustainable change, *Affordability 2.0* offers the insights and tools you need to succeed. If you are ready to challenge traditional thinking and embrace a new framework for achieving affordability, this book is for you.

Continuous Improvement Series
Series Editors: Elizabeth A. Cudney and Tina Kanti Agustiady

Additive Manufacturing Change Management: Best Practices
David M. Dietrich, Michael Kenworthy, and Elizabeth A. Cudney

Healthcare Affordability: Motivate People, Improve Processes, and Increase Performance
Paul Walter Odomirok, Sr.

Robust Quality: Powerful Integration of Data Science and Process Engineering
Rajesh Jugulum

Statistical Process Control: A Pragmatic Approach
Stephen Mundwiller

Transforming Organizations: One Process at a Time
Kathryn A. LeRoy

Continuous Improvement, Probability, and Statistics: Using Creative Hands-On Techniques
William Hooper

Affordability: Integrating Value, Customer, and Cost for Continuous Improvement
Paul Walter Odomirok, Sr.

Design for Six Sigma: A Practical Approach through Innovation
Elizabeth A. Cudney and Tina Kanti Agustiady

Building a Sustainable Lean Culture: An Implementation Guide
Tina Kanti Agustiady and Elizabeth A. Cudney

Lean Sustainability: A Pathway to a Circular Economy
Elizabeth A. Cudney, Sandra L. Furterer, Chad M. Laux, and Gaganpreet S. Hundal

A Career Worth Engineering: Don't Just Graduate-Navigate the Transition from Student to Professional
John S. Rogers and Sean Maciag

Lean-AI: A Humanistic Integration of Lean and AI
Adil Dalal and Elizabeth A. Cudney

Affordability 2.0: A How-To Roadmap for Application and Implementation
Paul Walter Odomirok, Sr.

For more information about this series, please visit: https://www.crcpress.com/Continuous-Improvement-Series/book-series/CONIMPSER

Affordability 2.0

A How-To Roadmap for Application and Implementation

Paul Walter Odomirok, Sr.

CRC Press
Taylor & Francis Group
Boca Raton London New York

CRC Press is an imprint of the
Taylor & Francis Group, an **informa** business

Designed cover image: Paul Walter Odomirok, Sr.

First edition published 2027
by CRC Press
2385 NW Executive Center Drive, Suite 320, Boca Raton FL 33431

and by CRC Press
4 Park Square, Milton Park, Abingdon, Oxon, OX14 4RN

CRC Press is an imprint of Taylor & Francis Group, LLC

ISBN: 978-1-041-30115-8 (hbk)
ISBN: 978-1-041-29991-2 (pbk)
ISBN: 978-1-003-77293-4 (ebk)

DOI: 10.1201/9781003772934

Typeset in Times
by KnowledgeWorks Global Ltd.

Contents

Preface

Albert Einstein once said, "Problems cannot be solved by thinking within the framework in which the problems were created." This profound insight serves as the foundation for the book you are about to explore. Affordability 2.0 is not merely a continuation of the ideas presented in my previous works; it is a transformative evolution—a new framework for thinking and solving complex problems that plague organizations, industries, and society at large.

Nearly a decade ago, I published *Affordability (Integrating Value, Customer, and Cost for Continuous Improvement)* in 2016, which introduced a structured approach to achieving a state of Affordability. Two years later, in 2018, I expanded on this concept with *Healthcare Affordability (Motivate People, Improve Processes, and Increase Performance)*, applying the Affordability framework to address the pressing challenges in healthcare. Both books were inspired by my experiences at Northrop Grumman, where the concept of Affordability first emerged during a project aimed at improving the quality, speed, and cost of the F/A-18 Super Hornet aircraft. That project, aptly named "Affordability," laid the groundwork for what has now become a comprehensive methodology for addressing complex problems.

Fast forward nearly ten years, and the need for Affordability has never been more urgent. Across the United States, issues such as Healthcare Affordability, Housing Affordability, Nutrition Affordability, Government Affordability, Military Affordability, and Education Affordability dominate public discourse. These challenges are not merely economic; they are cultural, systemic, and deeply ingrained in the way we think about value, cost, and performance. Affordability 2.0 offers a new framework for addressing these issues—one that moves beyond traditional thinking and embraces innovative approaches to problem-solving.

The traditional framework of thinking often focuses narrowly on price, assuming that lowering prices or providing goods and services for free will solve affordability challenges. However, such approaches frequently lead to unintended consequences, creating new problems rather than solving existing ones. For example, the Affordable Care Act, signed into law in 2010, aimed to make healthcare more affordable. Yet, 15 years later, it remains reliant on federal subsidies to maintain affordability for patients, highlighting the limitations of a price-centric approach. Affordability 2.0 challenges this mindset, emphasizing the importance of balancing value delivery, customer requirements, and cost considerations to achieve sustainable solutions.

At its core, Affordability 2.0 is about adopting a new way of thinking—one that integrates value, customer needs, and cost in a holistic manner. It is a framework for leaders and managers who seek to transform their organizations, improve performance, and achieve prosperity. This book provides a roadmap for implementing this framework, supported by real-world examples and proven tools, techniques, and methods.

About the Author

Paul Walter Odomirok, Sr. has spent the past 50 years building a diverse career spanning academia, engineering and development, manufacturing, corporate executive leadership, consulting, and entrepreneurship. In his 'first career', he taught mathematics and computer science at all levels of learning, from pre-school to post-PhD. He was even involved in designing graduate level curriculums at the University of South Carolina as an adjunct professor. His second career began in 1985 with NCR (National Cash Register). Beginning as a senior programmer analyst, he was promoted to manager—software development; product manager; manager—product management; corporate strategic planner for banking and retail; director—retail systems product integration; retail systems manufacturing; director of quality and customer satisfaction; and corporate coach. While at NCR, he was involved in the NCR/AT&T merger and was responsible, as a director of quality, for the cultural transformation and change leadership for the Retail Systems Division Lean Manufacturing Plant. Leveraging his involvement with Bell Labs Organization Effectiveness Group and Harvard Business School T.E.A.M. Research Study, he left NCR/AT&T in 1995 to pursue a consulting career, in the areas of leadership, strategy, and performance improvement. Over the past 30 years, he served 137 organizations encompassing over 300 projects and programs. Some of his clients include Anixter, AT&T, Boeing, Boston Whaler, Emory, Fibervisions, Georgia Hospitals, IISE, Motorola, GM, NASA, Northrop Grumman, Perkins and Will, U.S. Department of Defense and many major universities. Markets served include: engineering, manufacturing, healthcare, aerospace, telecommunications, government, defense, supply chain, energy, retail, banking, academics and higher learning. His first book, *Affordability: Integration of Value, Customer and Cost for Continuous Improvement* was released December 2016. His most recent book, *Healthcare Affordability: Motivate People, Improve Processes, and Increase Performance*, was released September 2018. He is currently, president and CEO of Performance Excellence Associates, Inc.

1 Affordability 2.0

Price is what you pay. Value is what you get.

– Warren Buffett

AFFORDABILITY AND THE CUSTOMER

Time and time again, the cost of living is an everyday topic for almost every adult in the United States. The cost of living involves the expense for shelter, clothing, food, energy, transportation, safety, taxes and pleasure. Most everyone, young and old alike, single individuals as well as couples and families, young adults, mature adults and senior citizens pay attention to their daily, monthly and yearly expenses. Although there are some exceptions, people, for the most part, are on the lookout for affordable solutions for their everyday personal costs. Organizations and groups are also in search of a better way to spend their money on their operating requirements. Whether one's marketplace is targeted at everyday consumers, or the business target market is dependent on service suppliers and/or product producers, the solution involves affordability, where the expense or spending is less than the revenue to be profitable. If operational affordability is a tenet, it's logical to ask, why not apply the theory of affordability to manage expenses, and deliver value to the customer, or use the theory of affordability to manage personal costs, expense and spend? To understand the entire Model of Affordability, and its systems for success, one must understand and be aware of the relationship of the three dimensions of Affordability. Affordability is based on three variables: Value, Customer and Cost. Combining those three factors for focus, the Aim of Affordability becomes a balance of those dynamics to achieve profitability and prosperity.

Affordability is not a destination nor a static target to pursue. It is a pursuit of an aim that exists between Value, Customer and Cost. With all that said, the question arises, "How do you know you have achieved some level of "Affordability"? The answer is clear: when the Value you deliver, meets and/or exceeds the requirements of the Customer, and the Customer can, and is willing to pay the price charged, while the revenue received covers all expenses with a profit margin. That definition may be clear, but it is not simple and easy to accomplish without an in-depth understanding of each primary component. To simplify even further, let's look at the three dynamics of Affordability:

1. **Value** is delivered to the Customer in a manner that meets and/or exceeds Customer requirements using resources, methods and tools at an expense or Cost.
2. **Customer** receives Value delivered per customer requirements, and pays a price the Customer can, and is willing to pay as a Cost.
3. **Cost** for delivering the Value is the expense of the resources required to deliver the Value, and cost for the Customer is the price the customer is willing to pay (revenue).

DOI: 10.1201/9781003772934-1

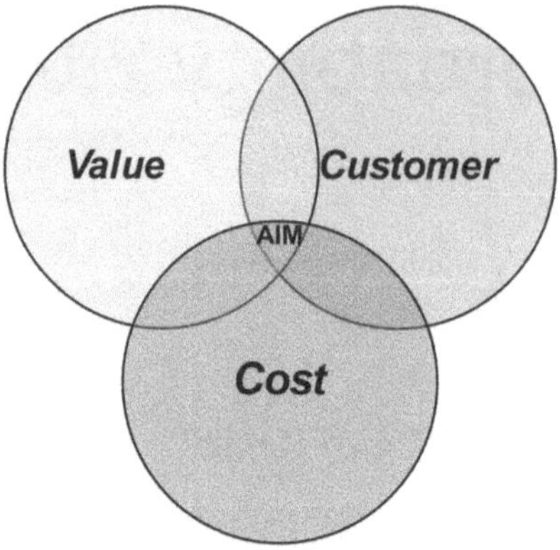

FIGURE 1.1 Affordability.

The **AIM** is the glue that holds it all together, the principles that guide, the methods and means of application, the functions that operate, the Shared Values, the Strategy, the Structure and the Systems that apply, and the content and construct of ***Affordability*** (Figure 1.1).

Affordability 2.0 is based on the original model with added features and functions that address the construct and foundation required to pursue Affordability. Over the past ten years, I've realized Affordability needs specific structure, systems and strategy to be implemented effectively. The original design of Affordability addressed: Customers, Value, Cost, Speed, Quality, Process, Performance, Change, Leadership, Creativity and Methods for Implementation. "Healthcare Affordability" (Published in 2018) leveraged the original Affordability model and vertically focused its content on: Strategy, Operations, Tactics, Solution Design, Implementation and Maintenance of Healthcare. Now, ten years later, "Affordability 2.0" has more clarity content through projects and programs of the recent past, as well as some others executed within the past 35 years.

THE CUSTOMERS' PERSPECTIVE

Affordability is based upon delivering Value to meet the requirements of the Customer. Each Customer has their own perception of Affordability that is dependent on their own economic means and their understanding of the marketplace. As a "Customer," I often consider Affordability in terms of the basic everyday essentials: gasoline, eggs, meat, mortgage/rent, transportation, meals, entertainment, etc. Since I don't own an E-Vehicle, the price of gas drives one of my perceptions of Affordability. Currently, at $2.70–$3.25, gas is Affordability for me. If it's less than $2.00 per gallon, I am elated; more than $3.25, I am dissatisfied. In California, $3.19 is very Affordable, since gas is over $5.00 per gallon. However, $6.00 or more than $7.00 may drive a perception of unaffordability (Note: California gas has been more than $7.00 within the past four years, nearly $6.00 as of the writing of this book.). Consider eggs ... for me, $3.00 or $4.00 per dozen is Affordable. Eggs for $5.00

per dozen or more is approaching unaffordability. For me, meat (beef) at $14.00 a pound or less (ground beef, roasts and steaks) is Affordable but $25.00 a pound is a little high. For me mortgage/rent for $1,000 per month may be Affordable, but for others, it is not possible. Since I'm debt free, no mortgage, no rent, I am definitely in an Affordable realm since it's free', but there was a price paid to get there. As for transportation, I don't have a car loan, and in an area that requires transportation, so my Affordability factor is positive. With restaurant meals in mind, $20.00, $30.00, or even $40.00 per meal is Affordable with a check for the two coming to less than $100.00. But for a family of 6, a dinner for $180.00 to $240.00 may not be Affordable. With this thinking in mind, what's Affordable for me, may not be Affordable for others, or what's Affordable for others may not be Affordable for me. For example, if the price of a Lamborghini is $233,000 to more than $500,000, it's not Affordable for me, but Affordable for that group of consumers who buy high end luxury automobiles. Every individual often has their own set of "Affordability Factors" for the products and services they purchase.

Affordability is mostly dependent on the individual Customer, and their ability to pay to receive Value to meet the requirements they have for a product or service they desire. Often, we hear the word affordability tossed around as if there's a specific price point or amount charged. Most often, Affordability is within a range of prices for customers who may fall in a particular target market. A Lamborghini is not Affordable for me, but the Jeep Rubicon is (which was my last automobile purchase in 2024). But one does not buy a vehicle on price and price alone. In addition to price, vehicles are most often purchased after careful consideration of the needs, wishes, wants and requirements. In my case, living on Anastasia Island, St Augustine Florida, I wanted a vehicle equipped for driving on the beach, pulling a fishing kayak trailer, traveling offroad when necessary, with the ability to be used for everyday activities and errands with all the interior features offered in most SUVs. The Jeep Rubicon met my use requirements at a price I was willing to pay, or in other words, for me, it was Affordable. Affordability occurs when a customer purchases a product or service that meets the customer's requirements at a price point that the customer can and is willing to pay. When the Value delivered by a product or service meets or exceeds the Customer's requirements, at a price the customer can and is willing to pay, Affordability is achieved.

This framework of thinking is different from the traditional pricing approach under the classic supply and demand model. In fact, within the construct of Affordability, demand is a goal to grow, and supply is a challenge to the system to control. By delivering Value that meets and exceeds customer expectation, the target market customer base grows in terms of demand. Supply is configured to be available, just in time, according to demand, and the flow of components for the product or service is coordinated with consumption. Price is set to be competitive at a level where the customer can and is willing to pay. Price is related to Cost considerations involving the expenses for the resources required to deliver Value. When revenue, from purchases are at price points meeting customers' expectations, and the revenue exceeds expenses incurred to deliver value, the organization is successful and stays in business. When the Affordability triumphant of Value, Customer and Cost is in synch, the organization is profitable and prosperous. Affordability 2.0 builds upon this foundation.

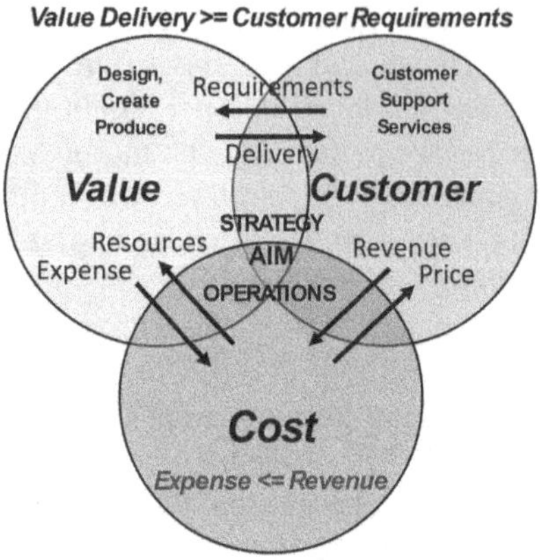

FIGURE 1.2 The dynamics of affordability.

Producers and providers must continually seek to aim and balance Value, Customer and Cost such that the Value meets and exceeds the Customer's requirements and expectations, the Customer is satisfied and delighted with the Value they receive, and the price the customer is willing to pay provides adequate revenue and margin to meet the expense of the resources required within the entire Cost structure. Profitability and prosperity is realized when the revenue received, from the price paid by the Customer exceeds the expense incurred to create, produce and deliver the Value. The Value Stream is the method and means used for Value Delivery. Figure 1.2 demonstrates the visual details:

WHEN THE SEEDS OF AFFORDABILITY WERE PLANTED

Fragments of thought can be developed to discover great concepts, theories, hypotheses and inventions. Those fragments can appear after events, thoughts, conditions and experiences. For me, the fragment of thought that seeded Affordability occurred almost 20 years ago on September 11, 2001. That morning, I was working on a client project, and shortly after 8:00AM, I went to the local Dunkin Donut to get two coffees and two bagels for me and my wife. Returning home, just before 9:00AM my wife told me that there was a plane crash in New York City in the north tower of the World Trade Center. As I watched the news report, another plane was flown into the south tower of the World Trade Center. Within about half an hour, a third plane crashed into the pentagon, and a fourth plane crashed in a field in Pennsylvania another half hour later. The rest of the day was a bit of a blur, full of confusion, that culminated with a speech from President Bush from the oval office at 8:30PM.

The next morning, I started the day developing some information and material for a local client I was working with regarding an organization transformation project. The customer called about mid-morning and explained that the project will be on hold for the "time being." Later that day, a few other organizations I had plans and commitments from for delivering my services, either delayed their projects, or cancelled the work for the rest of 2001. By the end of September, every one of my "promised" and pending projects were either "delayed" or cancelled for that year and the next. My business went from flourishing to floundering in just a few weeks. My financial outlook and revenue flow went from very good to zero.

As 2002 dawned, I didn't have a single project or program to work on. After three months of frustration, in January, I began the investigation of alternative sources of income. Old clients were still in a frozen or in an "on hold" status for leadership development, strategic planning, process improvement and people performance. New customers were nowhere to be found from my purview. I considered a variety of new business ventures.

By the middle of February, after several weeks of researching and considering various opportunities, I decided to open a neighborhood coffee shop that focused on serving the local community in which I lived. I was able to identify an ex-TCBY store that had just closed a few months ago. It was in the center of the community where several thousand homes were located, near one of the largest high schools in Georgia (about 4,000 students in 2002) and near the largest middle school in Georgia (about 3,100 students in 2002). The location was about five miles from two Starbuck's locations (one to the northeast and one to the southwest), nine miles from another similar coffee house in Duluth, Ga (to the northwest) and a couple of miles from that local Dunkin Donuts just to the north. Within a five-mile radius of my planned location, there were tens of thousands of adults and older teenagers.

After two months of purchasing the necessary equipment, signing a lease and obtaining county approval and licensing, I opened "Somethin's Brewin" on Monday April 15, 2002. As it is with most retail establishments, the first month was quite busy. The business operated six days a week, Monday thru Thursday 6:30AM–9:30PM, Friday 6:30AM–11:30PM and Saturday 8:00AM–11:30PM. My biggest challenge was differentiating myself from the competition. Dunkin Donut, at that time, had a limited food offering, as did Starbuck's. The nearest true coffee house was beyond five miles of my location, serving a different community where a natural boarder of interstate I-85 existed. I had to clearly define and determine all the specific requirements of the different customer groups I served. There were distinct groups and subgroups of customers. Men and women, teenagers, young adults, middle aged adults (with and without families), older adults and senior adults. There were several races and ethnicities in the area, and a diversity of gender and sexual preferences. A portion of the population were raised locally, with some people with a midwestern or northern background, yet others came from different areas of the south. A few customers came from the west coast, Hawaii and Alaska. And there were even customers from Europe, Asia, Africa, South America, Canada and the Pacific Islands, including Australia. The perception and expectations of a coffee house or coffee shop varied widely considering all the origins and dynamics of the customer base. In order to meet and exceed all the customers' requirements, wishes, wants and needs, I had to consider a wide range of design and implementation factors. I had to design and prepare the shop for operation. I had to determine the product offering in terms of food and drinks. I had to define the basic services and additional services that were vital to perform at a level of excellence and differentiate my coffee house from my competition in the area.

Preparation and Decorum: In order to set the Coffee House ambiance, I hired an artist to paint coffee house themed murals on the walls in the shop and even in the bathroom. I hired a carpenter to create the cabinets needed for the production area, a large four-sided kiosk for books and magazines, a coffee bar that was hung on a long wall and a long customer service countertop with storage cabinets below. The necessary equipment that was required included an elegant Italian Espresso machine, a quality coffee brewer from Fetco, an industrial blender from VitaMix, an ice machine, a computer system and numerous power outlets, a walk-in freezer, an industrial refrigerator, pantry shelves for canned goods and dry goods, an industrial oven, a microwave, a stainless steel food prep area, a stainless steel sink, an industrial dishwasher, grease traps per the county's requirements, a rest room, a self-service counter with trash receptacle, cups, glasses, plates, utensils, cutlery, as well as a multitude of other devices and gadgets. A retail transaction system for the financials and a standalone computer connected to the internet for customer use (Note: In 2002, public internet access available to customers was a relatively new service in retail establishments). The tables and chairs were sourced from a restaurant furniture supplier and placed in such a manner as to control customer flow into the shop, to the area for order placement, and through to the self-serve area where customers can then move out for takeout orders, or stay in for the inside experience. When it was finished, the design was unique, inviting and prepared to provide a pleasant experience for customers.

People: The people I initially employed ranged in age from 17 to 27, mostly local high school and college students. Over time, I hired retirees and middle-aged adults who demonstrated an interest in the business and frequented my shop as customers. The knowledge and enthusiasm of my employees was instrumental in growing the business over time. In the beginning, it was hard for me to explain their loyalty and dedication, until one of them said, "it's more of an experience and thrill to serve people who appreciate what you do." The staff were always eager to learn, pass the knowledge on to others, and strive for consistent standards and processes in order to serve high-quality, custom-made beverages and food offerings.

Products: The primary beverage, of course, was coffee bean drinks (Note: Coffee Beans are one of the top ten commodities traded worldwide). We brewed traditional coffees from arabica beans roasted by a local roaster to my specification at various levels from light to medium to dark. I was also able to source beans from various locations in the world and offer coffee products made from beans that Starbuck's didn't offer, or often, couldn't source in large quantities (e.g., at that time bean producers from the Tarrazu region of Costa Rica feared a Starbuck's takeover and resisted selling their beans to Starbuck's. Also, the limited availability of small Peaberry Beans from Costa Rica were uncommon to my competition. Large Maragogipe beans from Mexico became popular, as well as other unique varieties from Central America, Africa, Yemen, South America and the Pacific Islands). We created a wide range of Espresso based drinks such as straight espressos, lattes, cappuccinos, americanos, macchiatos, cortados, galaos, cubanos mostly in a hot presentation form, but also in iced and frozen concoctions. Custom coffee drink requests were always permitted and served per specific requirements. In addition, we offered teas, flavored Italian Sodas and a range of standard soft drinks plus "good old fashion filtered water." Breakfast food was available until 10:00AM (i.e., bagels and biscuits with a choice of egg, cheese, bacon, ham or chicken). For lunch or dinner, we offered sandwiches and croissants with a choice of a variety of proteins and vegetables (i.e., lettuce, pickle, tomato, onion) with options of a variety of condiments. Our biggest seller in the morning was a bacon, egg and cheese biscuit, and at lunch and dinner time a homemade chicken salad croissant with tomato and lettuce.

Services: The largest patron organizations for us consisted of schools and churches who had us cater meeting and gatherings. I entertained men's and women's Bible Studies, mostly in the mid-morning timeframe, with a local small church group holding service on Sunday ("my day off"). We held high tea for the Girl Scouts, "kaffe klatch" for the Boy Scouts (featuring caffeine free coffee) and participated in local community events, including 5K Runs, local school events and charity drives. Many of our services focused on children and teenagers which in turn attracted their parents, with an atmosphere and culture that provided a clean, safe and affordable environment. Since we couldn't have a drive-thru facility, we offered sidewalk grab-and-go coffee for commuters on weekday mornings. Lastly, we offered free coffee for community servants including first responders emergency personnel, local police, local fire and bus driver individuals.

Entertainment: To attract local citizens after dark, we offered different forms of entertainment on different nights. Tuesday was art night where a local artist led activities involving a variety of art venues. Wednesday was amateur night for locals, mostly students, to perform poetry readings, small plays, recitals and to display crafts such as woodworking, science experiments and needle point. Friday and Saturday were known as "Music Night" featuring music performed by local musicians. All in all, Friday and Saturday generated the most revenue for the week, as well as the most fatigue given the long hours.

After three years of operation, when it was all said and done, and in summary, more than anything, I asked myself then and even now; What did I learn? What was revealed? What did I really discover? What were the dynamics that carried for me those three years?

- First of all, the delivery of value is the driving force that maintains a business and generates customer loyalty, as well as sustains revenue flow to create profitability, achievement and success. The Delivery of Value must meet and exceed the Customer's requirements and expectations for satisfying demand, and being worth the money paid for the product and service.
- Meeting and exceeding customer requirements provides just the "ground floor" or bare minimum for prosperity. A greater success exists in the realization of "customer delight" that occurs when you go beyond the requirements, and venture into a realm that may not have been concisely defined in terms of all of their true "requirements, needs, wishes and wants."
- The price, or the cost of a product or service is a customer perception and competitive reality. It dwells within the customers' willingness to buy and their capability to purchase. It is established in the marketplace and must be met by the business. The ultimate goal is to establish a price where the value is considered "well worth it."

- There are both quantitative and qualitative factors that go into performance and the pursuit of excellence. Time, quality and cost are all relevant for successful delivery of a product or service. Satisfaction, even to the degree of delight, is a critical human factor for customers, employees and suppliers that impacts perception and assists in determining success and prosperity.
- Performance is the outcome of the system, and the more the processes are stable, standard and predictable, the greater the guarantee for delivery of value and performance excellence.
- Finally, and certainly not lastly, it's the people you employ in a product and service environment that drives the system and processes resulting in performance, profitability and prosperity.

In the summer of 2004, I began getting queries, requests and demands for my consulting services again. As the year progressed, and as more demand for my services materialized, I decided to sell the business and all the assets and "hang out my consulting shingle" once again to continue my pursuit of executing projects and delivering programs. As I studied and considered my next step, I strategically changed the name of my consulting corporation from "People Success, Inc." to "Performance Excellence Associates, Inc." I must admit, the "restaurant business" was a tough venture for me. Working six days a week, 6:30AM to 9:30PM Monday through Thursday, 6:30AM to 11:30PM on Friday and 8:00AM to 11:30PM on Saturday, totaling 94 to 106 hours a week was very exhausting. Although I "had enough" by the time 2004 was over, I realized that a three-year run from-scratch-to-profitability-to-sale is an uncommon achievement in the food service business. During the third year of operation, I was able to meet a level of profitability where my revenue greatly exceeded my costs and expenses without spending 100 hours a week. In open disclosure, I did not draw a salary during those three years of operation. And, looking at it now, I guess I could "chalk it all up" as an investment that earned me a PhD in Leadership, Strategy, Process Improvement and Performance Excellence from "S.H.D." ("The School of Hard Knocks"). Affordability 2.0 serves as my dissertation on the subject matter. And, not even recognizing it at the time, the learnings I gained that emerged from that venture, set the foundation, framework and course for my understanding of Affordability and my definition of Affordability. Subsequently, I was able to apply these principles to other organizations via various projects over the next several years.

THE STAGE WAS SET: 2005–2007 ORBIS AND ANIXTER

I started 2005 with two clients, committed projects and a sense of urgency. Orbis, a manufacturer of recyclable plastic containers, and Anixter, an electronic distribution company, both "invited me in" to their local sites to perform an assessment resulting in two different plans for action. In both cases, I visited each site for three days at a time on two different occasions. The resources available, and the options to pursue varied widely. The solution plan and design for each establishment differed a great deal, and the ability to act in each case was a day-and-night contrast. However, both organizations need a touch of Affordability.

The Value Delivery of the Orbis plant used structural foam injection presses that produced recyclable plastic pallets. The pallets are used by organizations like WalMart and Toyota to palletize and transport material. If the pallets were damaged or deemed unusable, they could be ground down into small plastic chips, which is in turn "foamed," injected into the presses thereby making another pallet out of the same material. I found it interesting that a plastic product, being recyclable, could be considered a "green" approach for reuse.

When I arrived at the plant, I noticed that at the end of the parking lot, there seemed to be a designated dumping area. All types of trash appeared to be separated into heaps. As I entered the plant, it was clear to me that there had been some type of machine failure because there was a lot of water and hydraulic fluid on the floor in, and around the press area. They had two presses, one in operation and one down. My guide, the director of quality for Orbis, and I met with the General Manager in an area just in front of the two molding machines. The inside of the building also appeared messy and ill kempt. As I "toured" the process, from incoming material to outgoing product, it was easy to observe, the plant was operating behind schedule in a mode of constant urgency, emergency and with virtual responsiveness to the breakdowns and system failures that plagued the plant. After a couple of visits, I was able to put together an assessment and plan for improvement with the assistance of the General Manager, Director of Quality and three of their most experienced operators. The initial emphasis of the plan used the "Theory of Short-Term Wins" (I often use the term "Quick Wins" interchangeably) to make some notable changes, in a short period of time to create successes and generate momentum for the improvement effort (Note: We were, in essence, picking off the low hanging fruit to motivate and stimulate the operators for process improvement). As successes were realized, I was subsequently invited in to some of their other plants in Kentucky, Ohio and Wisconsin.

In Kentucky, I was first exposed to one of their plants that created containers and conveyance products for the Toyota Camry factory in Georgetown Kentucky. It was literally located just outside the back gate of the Toyota site on land that Toyota provided. As it was with the first plant, I spent a few days on two separate occasions to assess and understand the challenges and "opportunities for improvement" that existed. As I did before, I assessed the situation to gain an understanding of the processes and systems at work, I gathered the facts and data that provided statistics and performance trends, and I took inventory of the resources and available options as well as likely targets for implementation of improvement efforts. With a plan and design in hand, using floor operators, we were able to make significant improvements in speed of production, quality of products to meet Toyota's demand, as well as lower expense and cost. Also in Kentucky, Bardstown specifically ("The Center of the Kentucky Bourbon Industry") another Orbis plant that made pallets to ship Kentucky Bourbon, had similar challenges, hurdles and roadblocks (i.e., quality issues, product availability, extensive budget overruns and employee frustration and discontent. Using the same pattern of approach: (1) assess, research, understand the data, facts and information,

(2) identify resources to utilize and options to pursue, (3) develop a solution plan and design, (4) pilot and experiment the solutions chosen, (5) test, evaluate and gauge the success until solution works, (6) implement the solution and make it permanent, (7) standardize and sustain the solution. Start first with quick wins to gain momentum and confidence with the workforce and continue working down the priority list.

In Ohio, I encountered their best performing plant. It didn't have a lot of waste, the material for production was readily available, the machines were well maintained and the manpower was motivated. The processes used in the plant were well documented, stable and predictable. It was commonly referred to as their best performer. After a couple of visits, it was clear to me that the use of Six Sigma was appropriate. The Ohio plant was hailed as the best in class for their operation. The General Manager of that plant was very savvy in terms of systems, processes and especially value delivery for every customer.

In Wisconsin, the Orbis plant was fairly near the Corporate Headquarters in Oconomowoc Wisconsin and was highly influenced by "the ivory tower." Although it operated somewhat efficiently, I realized it could use improvement. I employed the same consistent approach, using the General Manager, the Director of Quality and several machine operators to increase performance to realize a system that operated faster, performed better and produced more affordable products.

Anixter, on the other hand, being one of the largest electronic component distributors in the U.S., had some issues like Orbis, and in addition, unique challenges that commonly occur in the warehousing industry. As it is with all projects I'm involved with, I was invited in for Performance Excellence Associates Consulting. The site Quality Manager and the General Manager of the facility wanted the facility to better focus on process improvement and employee process quality capability. What started off as a local project, blossomed into a corporate wide effort for performance improvement.

Anixter Suwannee Georgia: A distribution center, within two miles of my office, provided electronic parts and components throughout the northeast Georgia region. After assessing the current condition, it was clear to me, the processes used required standardization and stabilization for optimum performance. The Quality Manager was a "Deming Advocate" and was quite eager to champion process quality improvement. Using the "Theory of Quick Wins," we leverage some "low hanging fruit" opportunities to establish momentum for the effort, and succeeded in piloting small projects that yielded a considerate number of substantial gains. Within a year, the word reached corporate, and I got invited in to participate in a major project engaged in the deployment of cell towers for Lucent Technologies.

Anixter Alsip Illinois: The plant, 40 miles south of headquarters by way of I-294, was solely dedicated to providing parts and components to Lucent Technologies cell tower construction sites in the upper mid-west. The labor force was comprised of Teamsters Union workers with an Anixter management team. Initially, the labor and management conditions were conflictual, but eventually evolved into a co-existent, and ultimately into a collaborative relationship by way of several small events by leveraging process improvements accomplished by union worker designed solutions. To begin the effort, the General Manager was replaced by an individual who embraced the Affordability Principles, proven Quality Improvement Techniques and the use of workers to improve their own workplace. Now, the management supported the effort, the workers embraced the approach, and with the improvements made, the system operated faster, and better, at a lower cost level. With the successes piling up, the corporate senior project manager that was assigned to the effort, along with the vice president in charge of the Lucent contract, made a decision to focus next on the Memphis Tennessee hub that served a great deal of the central US and Atlantic Coast.

Anixter Memphis Tennessee: This location was led by a General Manager who was quite astute regarding distribution quality. My Anixter colleague, the Senior Project Manager and I, were assigned to the Memphis location to assist the GM in achieving a Lucent Distribution Industry Certification after failing an initial assessment. Since I was familiar with the assessment from my days with NCR and AT&T Bell Labs, I had an advantage and unique perspective about the process. Much of the content mirrored the aspects of Affordability, and although, at the time, I did not have Affordability formally defined, I did have knowledge of the requirements, expectations and elements of the certification and certification process. The document used for evaluating the location contained a lot of similarities with the Baldrige Award framework. As we did with our last project, we started with the "low hanging fruit," prioritized and meticulously addresses each area of deficiency. Each project and activity was designed to be addressed by a small team of operators directly involved with the process in focus. A baseline of performance was measured, a potential solution and improvement was piloted, and when a "fix" was proven, it was instituted and standardized within the system. Within months, Anixter requested that Lucent re-evaluate the plant for certification. Not only did the site pass, but that location scored the highest result Lucent had ever had for that appraisal tool. The keys to success involved purpose, focus, leadership, teamwork and a consistent problem-solving process. The model that was created in Alsip and Memphis was deployed throughout all the sites of the Lucent contract.

Anixter St Petersburg Florida: After victories in Illinois and Tennessee, the SVP (Senior Vice President) of Operations, assigned the Senior Project Manager and me to one of the distribution and kitting operations in St. Petersburg Florida. At this spot, Anixter served Cisco (a network technology company) along with Solectron (an electronics product assembler and manufacturer) and Jabil (a circuit board engineering and manufacturing provider) for the provisioning or material required for Cisco's internet equipment implementation deployment in Japan. Anixter's role was to provide electronic components to Jabil, production parts to Solectron, and installation kits for Cisco that would all be packaged and sent to Japanese installation sites for distribution throughout the country's installation sites. Although the components for Solectron and Jabil flowed on time to demand without a hitch, the kits for installation were frequently behind schedule and flawed. The installation kits were comprised of items such as screws, nuts, bolts, brackets, additional hardware and installation instruction sets. Some kits consisted of four or five items, while others contained fourteen elements, and some orders were as small as 10 to 12 kits while others were as large as 50 or more kits. The work was highly variable and unpredictable in terms of the types of kits ordered and the size of the orders. It was a five day a week, one shift operation, operating on a make-to-order on demand basis.

Typically, Cisco dropped eight orders in the morning and eight orders in the afternoon. For each order, the kitting manager would receive a message on his computer when an order dropped, and he had to walk across the plant to pick up the order (about three minutes each way), set up the kitting line and walk back to the printer for printed kit instructions, and then again for the shipping and destination instructions. The kitting manager was the team lead and often was directly involved in kitting the parts and pieces of the larger orders. He had six employees available to kit orders. With the two major challenges being on time and flawed kits, the Kitting Manager, Senior Project Manager and I concocted solution teams from the kitting personnel for problem resolution.

For the defective kit issue, which amounted to wrong, missing, under quantity or over quantity of parts, the teams were able to research the past data of kit defects and realized some interesting facts and phenomena regarding kitting procedures. The Japanese customers expected perfection, with every kit populated with exactly the right contents, with exactly the right amounts (not to few and not too many items). After piloting several potential solutions, the kitting team came up with a self-defined kitting production design. Initially, the kits were built individually by each employee. The team decided to institute a "kitting line" approach, where each kit was "manufactured" on a flow line that involved several employees in sequence. Each employee had the responsibility of placing two items in the kit that was moved to their position and moved on to the next employee. Like an assembly line, the team populated the kit according to the kit configuration and their role in the kitting process. To eliminate fatigue, the kitting line colleagues would rotate their positions every 15 minutes, thereby reducing the probability of "weariness issues." In order to guarantee consistency and continuity, a photograph was placed in front of the "kitter" (i.e., a "kitter" is one who builds kits) that showed what the kit looks like when it arrives, and another photograph was placed in front of the kitter that illustrated what the kit should look like when it was passed on to the next kitter. Each kitter handled only two items, and the last kitter was the auditor who pulled out kits that were defective (a quality review and problem-solving event was held at the end of each kitting order completed). Since many kits contained sub kits of screws, nuts, bolts and washers, the right amount was critical when it was received by the customer (the Japanese installers required the RIGHT AMOUNT! Not too few, not too many pieces. Perfection!). The solution implemented was quite elegant. One of the employees had a brother who was a pharmacist in Tampa. Using a new framework of thinking, he likened the screws, nuts, bolts and washer to pills in a pharmacy. He sourced pill counting devices and sensitive scales from his brother for use when sub-assembly kits had to be pre-built, packaged and readied for the kitting orders. This inventive solution eliminated the defective procedures of the kitting process.

For the "on time issue," using a spaghetti diagram, the Kitting Manager realized that for every order that dropped he had to walk six minutes to get the order for kitting process set up, and spend six minutes of walking to gather the instructions, and another six minutes to get the destination and shipping labels, he consumed 18 minutes each time an order dropped. With typically eight orders in the morning and eight orders in the afternoon, he spent on average 288 minutes (more than half his day) walking for purpose of processing the orders that dropped. The team's solution was clear: put a printer at the point of use which was the desk of the Kitting Manager. The estimated cost for the solution (considering the right printer and the associated wiring and installation costs) amounted to approximately $4,000.00. Being a capital expense, the Senior Project Manager and I had to present the solution to the CFO and make a request for a CAR (Capital Appropriations Request). The CFO response was, "No. Because if we implemented there, we would be required to implement that solution at all the other kitting operations across the corporation which amounted to over $40,000." We escalated the request to the SVP (our "sponsor"), explaining that more than half the Kitting Manager's time was spent walking which wasted at least $18,000.00 annually (the Kitting Manager's salary was $36,000 per year). The SVP approved are request and explained to the CFO, that since kitting site facilities all had the same issue, investing $40,000.00 would eliminate a wasted cost of walking amounting to $180,000.00 (Note: the time recovered for Kitting Managers, was rediverted to other functions, such as "plant quality auditor" and "plant quality trainer.")

Each of the examples cited above involved complex problem solving, change and transformation for organizations of people, where a condition once existed such that the organization was somewhat successful, it operated far short of the customer's estimation of performance excellence. For both examples, there were several factors that clearly demonstrated and clarified the initial fragments that began to forge the underpinnings of the definition and characterization of Affordability:

- Customer Focus
- Value, Value Stream, Value Delivery
- Values, Purpose, Focus, Direction
- Leadership
- People and Teamwork
- Resources and Tools
- Design and Plan
- Complex Problem Solving

THE TIME HAD COME: 2007 NORTHROP GRUMMAN AND MRAP

After the terrorist attack on September 11, 2001, the beginning, the ramp up and the deployment of American military personnel to Afghanistan and Iraq occurred between 2001 and 2006. In 2006, the Taliban escalated their efforts in Afghanistan and additional personnel were surged to that country, and meanwhile, a Civil War erupted in Iraq that created more chaos and conflict in that war theatre. As 2007 dawned, most of the military equipment was heavily engaged in combat that required repair of much of the gear, or replacement and upgrade of rest. Combined with the condition that the major cause of death came from IEDs in roads, roadsides, buildings and even human suicide attacks, people and tools were well worn and in need of attention and solutions. As the death toll increased, and the equipment repair or replace requirement escalated, the military sought solutions to mitigate all challenges. In summary, some equipment needed to be replaced, and above all, warfighter lives needed to be protected from IEDs. My role began in early 2007, when I was invited in to participate in a program focused on F/A–18 Super Hornet Production, and later that year, I was also invited in to participate in the Mine Resistant Ambush Protected (MRAP) vehicle program to help with the #1 Priority of the American Military: Mitigation of IED Deaths.

The F/A–18 Super Hornet E, F, G Project: As the wars progressed, older F/A–18 models (A, B, C and D) needed to be retired and replaced by newer E, F and G models. A colleague of mine, Dr. Elizabeth Cudney and I were invited in to work with Northrop Grumman to increase the speed of production, improve the quality of the product and reduce the cost that would ultimately create the ability to reduce the price for the customer (i.e., The US Navy and the USMC were the customers for those aircraft). The jet was, by government contract, a Boeing product, where 60% of the statement of work was the responsibility of Northrop Grumman. In 2006, 42 F/A–18's were produced and delivered to the Boeing plant in St Louis Missouri for final integration and subsequent sale to the US Navy and US Marine Corps customers. The rate of replacement showed that within a few years, 67 per year would be needed to replace the older models being retired. With the leadership of the newly appointed VP of Defense Aircraft, Mr. George Vardoulakis and Senior Executive Mr. Dave Armbruster, Dr. Elizabeth Cudney and I created a program that used Lean and Six Sigma principles to improve the speed of production, improve the quality of output and reduce the operating cost which would thereby enable Northrop Grumman to reduce the cost and ultimately permit Boeing to lower the price of the product for the US Government. After three years of implementation (i.e., April 2007–August 2010), Dr. Ashton Carter, Under Secretary of Defense (who reported to Mr. Robert M Gates, Secretary of Defense) wrote in a memo to all Government Military Acquisition Personnel titled: "SUBJECT: Better Buying Power: Guidance for Obtaining Greater Efficiency and Productivity in Defense Spending",

> The NAVY, for example, recently concluded negotiations for a multi-year procurement of 124 F/A–18 strike fighter and E/A–18 electronic attack aircraft, which will yield over $600 million (greater than 10 percent) savings to the Department and the taxpayer. The F–18 program was able to drive down cost for each lot of aircraft procured in the framework of a fixed-price incentive contract that meets Department's objectives for realistic costs, reasonable profit, a 50/50 shareline, and a 120 percent ceiling
>
> **(Page 7, Paragraph 3).**

The Northrop Grumman leadership team named the effort: "Affordability" (Hence the "official start" of my pursuit of "Affordability." Including: the title, the basic definition, the framework, the foundation and the contents of the concept. Subsequently, the outcome launched the two first texts and now, the final publication on the subject.).

The Mine Resistant Ambush Protected (MRAP) Program: In 2006, escalation of fighting began to occur in both Afghanistan and Iraq, and the death toll from IEDs started to rise at an exponential rate. Mr. Robert M Gates, Secretary of Defense at that time, stood up in front of congress in October that year, and made a request for funding to purchase and deploy MRAP vehicles to both war theatres to protect and save lives of American Warfighters. I was invited in during the month of August of 2007 until March of 2014, to be involved in, and address product production, integration, deployment, strategic planning, fielding, maintenance, sustainment, storage and program closure topics. Over a period of seven years, and nine distinct projects, my participation with the teams that tackled these challenges gave me additional insight into the Theory of Affordability and the elements that comprised its body of knowledge. In three years, at least 25,700 vehicles were produced and deployed (Note: Fastest fielding of a ground combat system in the history of the world!). During both wars, 10's of 1000's of lives were saved by the use and utilization of MRAP vehicles. Upon completion of the program, the US Defense Department learned new ways of identifying, developing, producing and fielding military systems in a manner much faster than the standard acquisition process (Note: Which often took ten years or more.).

Upon completion, several factors demonstrated the framework and foundation for thinking in terms of Affordability. Whether it be the creation, implementation and operation of a food and beverage retail establishment (Somthin's Brewin'), or the production of recyclable plastic pallets and containers (Orbis), or distribution of electronic components (Anixter), or production of defense aircraft (the Northrop Grumman F/A–18 Super Hornet project), or deployment of mine resistant vehicles to save lives (the MRAP Program), all projects and programs illustrated and demonstrated the same pattern of conditions and dynamics for success. The design center for Affordability did not focus on the money and money alone, but it did contain the following facets, always keeping the money in mind as a lagging measure and the ultimate proof of profitability and prosperity.

- Value, Values, Sense of Urgency, Purpose, Focus, and Direction.
- Delivery of Value, Meeting and Exceeding Customer Requirements, charging a Price the Customer can, and is willing to pay for, and operating at an expense and spending rate that does not exceed revenue. Of course this seems like common sense, and it is, but it typically is not in use for many businesses and enterprises I've encountered since 1995.
- Leadership of People and the Management of all the other "things."
- People and Human Factors in terms of purpose and culture.
- Proven Resources and Tools for improving and maintaining systems and processes.
- Solution Design and Implementation Plan that increased the probability of success.
- Solving Complex Problems using a Standard approach for addressing difficult challenges and issues.

THE PATHWAY EMERGED, THE JOURNEY PROGRESSED AND THE DESTINATION WAS DEFINED

Clearly, from the many examples provided earlier in this chapter, it doesn't matter if you're a small business, a medium size business, a large business or a government entity, Affordability can be applied, and can yield results in the areas of growth, success and sustainability. The first book, *Affordability* was written and published about ten years ago, and it contains tools and techniques for implementing Affordability. About eight years ago, the book *Healthcare Affordability* and it provides ways and means for instituting Affordability in Healthcare establishments.

Affordability 2.0 provides additional details and information beyond what "Affordability" and "Healthcare Affordability" offered for establishing, achieving and sustaining an effort of Affordability in any setting. Always remember, Affordability is not about offering products and services free or cheap. It is about delivering value to a customer that meets and/or exceeds requirements, at a price the Customer can, and is willing to pay, and operating an organization at an expense level and spending rate where revenue is greater than overall cost. Affordability is based on the foundation and framework of being a profitable and prosperous operation, with a continuous effort to improve systems and processes, and be robust in the ongoing effort to change and transform as required by the customer base, using the strengths and capabilities of the organization. Sustainability, profitability, prosperity and growth are all proven outcomes and successful results.

YOUR PATHWAY AND YOUR JOURNEY

Seek first to understand. Before even beginning the journey, as complete an understanding as possible is necessary before even stepping onto the pathway. Understand Affordability, understand "the customer," understand the value you'll deliver, understand how you will create that value, understand the people you'll need, understand the required resources and tools, understand the plan and design for success, and don't forget what you'll be up against in terms of challenges and competition. The key advice here is to understand as much as possible from a historical perspective, a current state reality and an emerging future (if that is even imaginable) before even taking the first step onto Your Pathway and Your Journey.

From your understanding, identify options you may pursue and resources you may need. Options should be determined by the Value you plan to provide and the Value that you plan to Deliver to the Customer. Resources run the gamut from people, processes and performance to methods, machines, materials and money. Create potential solutions for your vision that can be developed and evaluated. Determine which solution is best.

Design and plan for the implementation you chosen potential solution. Include as much detail as possible that completely describes your unique value proposition (UVP). Test and evaluate the solution. Enhance it if necessary and test and evaluate it again, and again, until it's ready for implementation. When your solution proves to be ready, package the result and prepare it for implementation.

Implement your solution, maintain it and continuously improve it. Use all of the Affordability factors, tools, toolboxes and techniques to develop, improve and grow your Value, increase demand, improve profitability and delight the Customer.

2 The Goal

> The goal is not to improve one measurement in isolation. The goal is to reduce operational expenses and reduce inventories and increase throughput simultaneously.
>
> **– Eliyahu M Gollratt**

THE GOAL(S) OF AFFORDABILITY

The goal of any Affordability effort is about delighting Customers, satisfying employees and pleasing suppliers, while increasing speed, improving quality and reducing cost. Goals encompass a beginning to end strategic focus, started with delivering Value, and utilizing an efficient and effective Value Stream, and ended with a financial transaction that produces revenue, attains profitability and generates prosperity. The goal is not to just make money. The long term expected goal is to increase demand, grow the organization and attain an ethos of sustainment.

Let's begin with 'First Things First'. There's no sense in beginning any effort unless there is a sense of urgency to act, a compelling reason for action, a vision to pursue or a clear purpose. Typical goals are in terms of growth and an increase in market share, or revenue, or profit or prosperity. What if 'The Goal' is defined in terms of Affordability? In other words, let's deliver a product or service that meets and even exceeds the customer's expectations, requirements, needs, wants and wishes at a price the customer can afford, and is willing to pay, all the while operating at a cost that yields profit?

What if I told you, "You can get a 10:1 ROI with a program that will improve your quality, increase your speed and accessibility all while reducing your cost and increasing your profitability and prosperity." You might respond, "How is that even possible?" For too many years, business managers have suffered from the illogic of tradeoffs. For example, the belief is, if you want a faster, more responsive organization, it'll cost you. Or, if you want to improve quality, it'll cost you. Or even, if you want to increase Customer satisfaction, it'll cost you. However, thinking with an Affordability mindset, performance excellence can be met using proven methods and tools, while decreasing cost, spend and expense.

In today's economic society, the overarching goal is often accumulating wealth. Wealth occurs as the outcome of successful systems of an enterprise that delivers value within the realm of Customer Affordability. Toyota is an excellent example of an organization that has achieved this status in a 70-year period (Tesla is a more recent example that has appeared in the last 20 years). Whether the example is Toyota or Tesla, the overarching goal is comprised of steadfast; Purpose, Principles and Values, Product and/or Service Value, Customer, Leadership, People, Suppliers, Strategy, Systems, Structure and Performance Excellence.

DOI: 10.1201/9781003772934-2

- Purpose: The reason for being. The driving force for action. The center of all actions and activity for Value Delivery. I discovered my purpose is to provide personal service that increases customer and client performance. My purpose is also to focus on both the personal and organizational levels for meeting my Customer's purpose. My Customers' purposes have come from a wide range of aims from saving lives to delivering quality products and services to past successes to transforming faltering organizations to pursuing greater levels of performance.
- Principles and Values: Of all the principles and values I've encountered for the past 30 years, there are three that seem to be a reoccurring theme with most organizations: Respect, Integrity, Trust. Of course, all organizations operate with specific principles both written and unwritten, published and implied, announced and unannounced. Most of my personal experiences and encounters involved organizations and people with primarily honest, trustworthy, commendable values and principles. Every so often people with ill intent and malevolent purpose emerge to serve some function of chaos or resistance. Fortunately for me, 99% of the time, people are for a good decent cause.
- Product and/or Service Value: At the heart of Affordability is the Delivery of Value. The Value defined by a product and/or service meeting or exceeding Customer requirements, wants, wishes, needs and even dreams. The target of the delivery of value should be customers who are able and are willing to pay the amount requested (Note: Which should be set at a competitive level).
- Customer: The one who purchases the value you deliver. It is sometimes the Customers' customers who receive the ultimate value. For example, Car Dealers are the Customers of Toyota, while their customers are the ultimate recipients of the value delivered. At times, one might have to consider the customer of the Customer for the requirements, needs, wishes and wants that describe necessary Value.
- Leadership: Discussed in detail in the next chapter, leadership is the individual or collaborative people governing the organization providing control, guidance and management of the organization. Setting direction, aligning resources and motivating the people are the necessary actions necessary for any group to be successful.
- People: Persons in the organization that either create and deliver value or support the creation and delivery of the organization's value. "People" is the topic of a later chapter, and the fuel for driving processes, realizing performance and earning profitability and prosperity.
- Suppliers: The strength of the enterprise and supplier relationship, often overlooked by many organizations, is critical to success using the Affordability paradigm. Years ago, a book was published titled, "Happy Cows Give Good Milk." The intent was primarily focused on employees, but I've found that the concept also applies to suppliers. One supplier in the Value Stream can disrupt smooth and efficient flow. A supply line can be shut down at great cost to a business. This happened to Toyota in

1997 when a fire in a supplier's plant impacted their 14,000 cars per day. A $5.00 valve produced by the Aisin company shut down all of Toyota's Japanese production when the plant had to close. Whether supply chain related, or value delivery associated, or service connected, suppliers are a key link to success and should be appreciated as a partner or colleague for success.

- Strategy: A plan of action, both short term and long term, to achieve an overall aim. The aim of Affordability is to deliver value that meets a customer's needs at a price the customer can, and is willing to pay, while operating at a spending level that yields profitability.
- Systems: The integration of processes within an enterprise that delivers the organization's value in the form of a product or service or both. Examples of such systems include engineering, procurement, production, delivery, sales, marketing, customer support, maintenance, financial and system support.
- Structure: The arrangement and composition of the activities directed to achieve the goals of the organization with well-defined roles and responsibilities, using an integrated design for alignment and communication
- Performance Excellence: A level of system execution that far exceeds competitive benchmarks and industry norms. Both Quantitative Measures (i.e., Time, Quality, Cost) and Qualitative Measures (Customers, Employees, Suppliers) should be used to determine the functioning level of the operation. A Strategic Scorecard of Enterprise metrics can be used to determine execution (i.e., Customer, Business, Process, People).

Strategic goals should be linked to operational objectives, and those objectives should be linked to targeted tactical metrics and measures which eventually link in all individuals for performance excellence. Ultimately, all of the people in the organization should be linked in to the tactical, operational objectives and strategic goals.

For any goal to be achieved, and for success to be fulfilled, a scoreboard or scorecard must be present to convey the current operational status. The question "How do you know if you're winning or losing if you don't know the score?" is common when evaluating performance excellence. In 1996, the year after I started my consulting business, Robert S Kaplin and David P Norton published the book "Balanced Scorecard" from the results of the work they had done beginning in 1990. After evolving their original work, Goals for Affordability can be described as a balanced scorecard in the areas of: Customer, Business, Process, People.

Customer Goals should be set and measured from the customer's perspective (Goals such as; Satisfaction, Ontime, Delivery, Price). Business Goals should be established from both traditional financial metrics and current state metrics measuring cost reduction and expense avoidance, as well as market share growth and improvement of value delivery. Processes can be measured from a time, quality and cost viewpoint for system speed, accuracy, excellence and frugality. Finally, People Goals can be created for understanding satisfaction, capability development, advancement and growth. Even when starting a small business, like my own, Customer, Business, Process and People Goals should be known before entering into a venture.

Learning from my time with NCR, Bell Labs and my exposure to Harvard Business School and Dr. W Edwards Deming, I started my business with the goal to provide consulting services that delivered solutions using complex problem-solving methods to increase the success of my clients, hence I named the company: "People Success Inc." I concentrated on two strategic areas: Leadership and Strategy, and Process Improvement. After being in business for ten years, I refocused my organization on enterprise- and system-level solutions (still applying my expertise to leadership and strategy, and systems and process improvement), and I changed the business name to "Performance Excellence, Inc." with enhanced overarching goals to further develop Leadership and Strategy for organizations, and achieve system level Performance Excellence for Customers and clients. I have held that concentration for the past 20 years.

Learning from Dr. Deming, whom I had only personally encountered a few times, that if a product and/or a service is of Great Value for the Customer, or clearly worth the price and expense, neither selling nor marketing costs are necessary for success. Without any selling and marketing expenses, the overhead burden is lowered a great deal, and profitability is increased. I also learned that lower overhead costs lead to greater profits, so, like Dr. Deming, I operated from a fully established home office (Note: He never had a brick-and-mortar office. The Deming Institute was a virtual organization with no physical location.). My initial overarching financial goal was to offer my consulting services at a value level worth paying an amount that was between the lower end and higher end of the price range of the industry without the overhead of marketing and selling expenses. From my first day, September 15, 1995, to this very day, I have not spent a dime on marketing or selling programs. In addition, the price point for my services ranges below the market's upper end, and above the market's lower end. My personal, or people goals and objectives for satisfaction, development and growth have continued (i.e., I've stocked a personnel library with over 800 books and I've read and used them all.). I've been certified by the IISE as a Lean Six Sigma Master Black Belt Sensei, my expertise has expanded throughout several countries and continents and I've delivered programs to 20 Universities in the US and Canada. I've learned that success and continued growth over time is a net profit payoff that is a pursuit everyone should consider.

Learning from other mentors and coaches, my services focused on areas of my experience, knowledge and credibility. From my corporate background with NCR and AT&T's Bell Labs, my academic background in Math, and my athletic experience in playing and coaching various sports, I was able to establish "People Success, Inc." concentrating on developing leaders, creating strategy and improving processes (using methods acquired in Quality Assurance, Lean and Six Sigma). In my tenth year, I shifted focus and re-named my institution "Performance Excellence Associates, Inc." with a continued concentration of leadership and strategy, and an increased focus on enterprise and system performance.

Today, more than 30 years later, my margins remain high, my overhead is low and my profitability is extraordinary. People often wonder how I was able to exist for more than 30 years without a marketing or selling faction, serving 137 clients through more than 300 projects, at a competitive rate. My back story is short and simple. My first three projects were handed to me by my first client. Clients and projects after 1996 were attained through references and referrals by my first client and subsequent clients. Professional relationships over the years have afforded me the luxury of having new clients appear by way of an old client or through recommendations of colleagues, associates and partners. Today, although I've recently reduced my consulting activity, my work comes from demand, request and repeat business only. The satisfaction and loyalty of clients sustain the business. During the past 20 years of delivering programs for the IISE, to literally 1000's of participants, my rating from program attendees has accumulated an average of 4.85 out of 5.00. My advice is to set your goals in the areas of Customers, Business, Process, People and monitor your qualitative goals of satisfaction and delight, and manage your quantitative goals of time, quality and cost.

AN IMPOSSIBLE GOAL

The MRAP program I mentioned earlier in this book, was the number one priority for the US military in Afghanistan and Iraq in 2007. Since IEDs were the main cause of death in the war theatre, the Secretary of Defense Robert M Gates, approached congress in 2006 requesting funding to field MRAPs to protect and save lives. Congress approved funding and the MRAP Program was "Stood Up" in January Of 2007.

I was "invited in" to participate in the program August 2007. My first assignment was on the SPAWAR Naval Base in Charleston South Carolina where they were preparing and integrating all MRAP vehicles for use in the middle east campaigns. My first day was August 22nd, and after passing through security, I was escorted to the large facility that contained all equipment and personnel focused on installation, integration and readiness for the vehicles destined to be flown to Kuwait for 'deprocessing' and deployment. My first impression can be summed up in two words: "urgency and chaos." Most of the workforce was comprised of mechanics, machinists, metal workers, electricians and equipment specialists who had recently retired from the military. Most of the workers there had friends and acquaintances fighting on the war front. The flurried activity seemed to have no rhythm, and the facility itself was cluttered with crates, empty and full, packing material, banding, wire, parts, pieces and equipment being prepared for installation. To sum it up, the place was a chaotic mess. That day, five MRAPs were completed and transported to the Charleston Airport for flights to Kuwait. I was told that Secretary Gates promised that 25,700 vehicles would be deployed in a matter of three years. Using the Takt Time Formula, I calculated the required rate of flow necessary to meet the commitment.

The Goal for the speed of integrating MRAPs can be established via the Takt Time Formula;

- A Work Day: 2- to 8-hour shifts (1/2-hour lunch, 2- to 15-minute breaks) → 14 hrs daily
- Total Work: 14 Hours a Day, 6 Days per Week, ~50 Weeks per Year, for 3 Years
- Total Time: $(14 \times 6) \times 50 \times 3 = 12{,}600$ Hours of Work
- Total Demand: 25,700 Trucks

- Takt Time (The Rhythm in Time required to meet Demand) = Total Time/ Total Demand
  Calculation: 12,600 Hours/25,700 Trucks = 0.49 Hours/1 Truck
  Daily Rate: 29.4 mins/1 truck, or ~ ½ hr per truck → 28 trucks per day

On August 22, 2007, my first day on site, five trucks were completed, placed in a military aircraft and flown to Kuwait. At the current rate of five trucks per day, using approximately 300 working days a year, 1,500 trucks would be completed in a year (Note: five trucks in one day were their best performance to date that year.). For 25,700 trucks, at the rate of 5 per day, it would have taken more than 17 years to complete all 25,700 trucks (i.e., 25,700/1,500 = 17.13). In other words, with only 276 trucks completed by August 22, 2007, at the rate of 5 per day, the last truck would have been completed in 2024.

By the end of the month, a new goal was set, and a team was assembled to lead the effort to increase the speed, improve the quality and reduce the waste of personnel, equipment, tools and materials (Figure 2.1). The new goal of "50 a Day" was established to overcome the current shortfall and put the organization on a pace to meet the commitment. This new goal was achievable since there were 25 pairs of two position service bays that could be servicing 50 trucks on each shift. If the first shift completed 25 trucks, and the second shift completed 20 trucks, 50 trucks were completed that day (Note: It required that the rhythm of the work output 25 trucks each shift while servicing 50 trucks simultaneously in the two-position bay configuration.).

During September and October, Lean Six Sigma methods and techniques were used for process and system improvement. My prediction was October 31st for the first 50 per day production run. It actually occurred on November 4th, and January 2008 was the first month where the average 50 per day rate was realized. From that month forward, the average 50 per day rhythm was established, and the 25,700 goal in three years was accomplished. What was an impossible goal to begin with

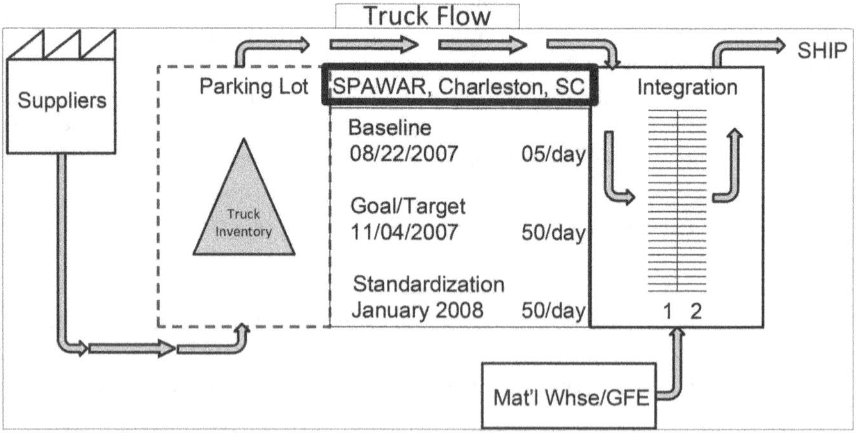

FIGURE 2.1 Truck flow diagram.

became an achievement that had never been experienced before. It was the fastest fielding of a ground system in history. Proving that even government, when focused and dedicated, with clear vision and direction, with forthright principles and values, can make the impossible possible. Considering the Goals of Customer Satisfaction, Worker Satisfaction, Supplier Satisfaction, Time/Speed of 50 per Day, Quality of Lives Saved, Cost of Waste Reduced, all primary Affordability Objectives were met.

WHEN JUST-IN-TIME WASN'T

My very first taste of Big Goals gone awry came in 1990. While working for NCR, I was transferred to NCR's new Retail System's plant in the summer of 1989. The plant was a "green field site" for NCR's new Retail System's assembly facility based on Toyota's Just-In-Time manufacturing principles and methods. The assembly line began operation in October of 1989, and its first full year of production was 1990. By the end of that year, the primary goal of Just-In-Time was measured at 60.2% Ontime-to-Commitment (Note: It wasn't on time to customer demand and request, but on time to the plant's commitment). The Just-In-Time factory was actually on time less than 2/3rds of the time. In fact, all the major goal areas (i.e., Delivery, Quality, Customer Satisfaction, Employee Satisfaction, Capacity, Revenue, Profit) all fell short of the goals established. The baselines established in 1990 served as the start-up for a five-year program focused on reaching a state of performance excellence.

As 1991 dawned, rumors of an AT&T acquisition of NCR began rolling around. At that time, NCR was the fifth largest manufacturer of computers in the world, and AT&T wanted to acquire the company and institute them as their own computer arm. Although most of NCR's computer products served the Banking and Retail industries, AT&T CEO and Chairman Bob Allen had made a commitment to the board that such an acquisition was strategically necessary and the takeover of NCR occurred in September of 1991 (Note: AT&T subsequently spun off NCR in the 1995 AT&T "*trivestiture*" that established AT&T Systems, Lucent Technologies and a once again independent NCR). During 1991, I was promoted to Director of Quality and Customer Satisfaction responsible for leading an effort to transform the organization and attain a greater echelon of performance.

Most of 1991 was dedicated to process improvement, eliminating defects and increasing speed. The value stream was comprised of eight primary processes (i.e., receiving, order processing, SMT board production, system assembly, scanner assembly, peripheral assembly, order consolidation and pack and ship) where scoreboards were displayed illustrating team performance in terms of people, process and performance. Every Value Stream Process had defined internal supplier inputs and distinct internal customer outputs (e.g., order consolidation was the "internal supplier" for pack and ship, and the package shipper was its "internal supplier.")

Measurements were taken for on-time delivery (daily), process quality (daily), "customer" satisfaction (monthly), process employee satisfaction (monthly), waste elimination and cost reduction. Each "self-directed work team" was responsible for their own budget and their own performance. This was the first example of a self-directed work team design that I was exposed to (Note: This was a two-shift operation with only one floor manager per shift to resolve conflicts and issues between teams and on the Value Stream.).

During 1992, an increased effort was placed on Out-of-Box Quality (i.e., Out-of-Box Quality consists of Quality, as it is received at the Customer site). Since we performed an eight-hour 'run in' of each product produced, our operational quality was at 100% when the product left the plant. The Out-of-Box ("OOB") Quality issues occurred with items missing from the orders (e.g., installation paraphernalia and incorrect components), wrong items delivered (e.g., cables and connectors), elements damaged during shipment, etc. In order to assure quality, we instituted Field Quality Audit (FQA) where we sent a team of Associates (i.e., usually one engineer, one manufacturing employee and one other plant associate) to observe and film the delivery and installation of our equipment. Each Quarter of 1992, we held a brief plant meeting to play the results of each FQA for that Quarter (Note: This is very effective for both the Associates and the Customers to understand the quality of the product.). As the year progressed, we worked 'upstream' in the organization establishing performance boards in groups that supported the Value Stream (e.g., Finance, HR, Engineering, Design, Software Development, Technical Support, Maintenance, Marketing, Sales, etc.). In 1992, we used the Baldrige Award Criteria to assess our performance on an overall basis. We scored 392 out of 1,000 and it created further stimuli and sense of urgency for the plant management. The sense of urgency to improve intensified since AT&T, our current parent company at that time, had won two Malcolm Baldrige National Quality Awards. We were now moving at an accelerated pace.

In 1993, after making good progress during the two previous years, we expanded our scope of the FQAs, extended our reach upstream to our suppliers and partners, and increased our emphasis on Customer Satisfaction by scheduling regular Customer Plant visits for all of our major customers. The plant visit was standardized, and after an introduction to the plant, and overview of all of our products and services, we would take the customer out to the production floor to see the products being produced 'live'. Most often, one of the Customer's products would be in production while they were taking the tour. By the end of the year, we were averaging 'a customer tour a day'.

Also, during the last quarter of 1993, we were able to leverage our relationship with AT&T to produce an eight-minute music video that featured all that had been done, and all that we were doing at "E & M Atlanta" (i.e., The Official Facility Title). The AT&T Video Production Team in Atlanta was in their slow period for the year, and they offered to come out and produce a video targeted at capturing what was happening at the plant. After a few weeks of design, and after a few weeks of shooting, "Our Common Bond" Video was ready for release and distribution. At the 1994 kickoff meeting held for all employees, the video was shown and handed out in '8-track' format to all Associates as they left the meeting. Since one video goal was to include all 1,000 employees, it was a hit for everybody since they each had a part in a music video (Note: MTV was quite popular in the early 90's and the plant's employees were proud to participate and show their family the music video that they were featured in!).

Also, beginning on January 1, 1994, we announced an employee bonus program based on "Performance-to-Goal" results and outcomes. In addition to the debut of the "Our Common Bond" Video at the annual kick-off meeting, the bonus program was announced. It was presented as follows:

> If all performance goals are met in 1994, everyone will receive 3% of their salary. If the goals are exceeded by 10% or more, everyone will get 6% of their 1994 salary. However, if performance is at 80% or lower as compared to the goal, 0% will be awarded.

A master display of the Goals (Note: Including the baselines and current state measures) was displayed in the main hallway that ran through the length of the building and began at the employee entrance. The hallway was architected with glass walls on one side that permitted viewing of the actual Value Stream (Manufacturing) and on the other side was the location of The Value Creation Functions (Engineering), with the rest of the plant resources (e.g., finance, human resources, security, maintenance, food service, support services, etc.). Each Display was 8' high and 4' wide, and each of the five performance categories had their own display: Customer, Associate, Delivery, Quality, Business. The Customer Goal (i.e., Customer Satisfaction 25%) and the Associate Goal (i.e., Associate Satisfaction 25%) made up 50% of the overall bonus calculation, while the Delivery Goal (i.e., On-Time Delivery 20%), the Quality Goal (i.e., OOB Quality 20%) and the Business Goal (i.e., Revenue, Expense, Cost Reduction and Profit Increase 10%) made up the other 50%. The Performance Goal displays were monitored and managed daily, monthly and quarterly depending on the ability to measure each goal area (i.e., Customer Satisfaction Quarterly, Associate Satisfaction Monthly, Ontime Delivery Daily, Quality Daily, Business Monthly.).

As 1994 ended, we found ourselves in a celebratory reality as a result of the five years of concentrated effort focused on meeting and exceeding the goals established in the categories that were originally created in 1989 when the plant opened. On the first Friday, January 6th 1995, each non-management employee eligible for the performance bonus was greeted at the door and handed a check for 3.04% of their 1994 salary as part of the bonus program put in place to reward employees according to how well the plant performed to the goal set. The 1994 results showed that the plant, on average, performed slightly above goal when all categories were combined. The median salary for non-bonus personnel was a little more than $30,000.00, so the bonus checks averaged a bit more than $1,000.00. It was quite a nice reward that extended the holidays at least a few extra weeks. To summarize, below are the 1990 Baseline Measures compared to the 1994 results:

Actual Goal Measure	1990 Baseline	1994 Outcome
Delivery Performance (Just-In-Time)	60.2%	92%
Overall (Out of Box) System Quality	87.0%	99.2%
Customer Satisfaction	72.1%	85.6%
Associate Satisfaction	52.0%	65.5%
Capacity	72,000	98,000
Product Revenue	$220MM	$262MM
Profit (after all expenses and taxes)	–$2MM	$5MM

So all-in-all, you might ask, how much did the bonus program cost in the end? With about 800 non-bonus eligible associates in the plant, getting a check of around $1,000.00, the payout was about $800,000.00 for a dramatic improvement in delivery performance (53%), an increase in out of box quality (15%), a 19% increase in customer satisfaction, a 26% increase in associate satisfaction, greater capacity, more revenue and lower cost for a profit-after-tax of $5,000,000.00 (Note: a $7M turnaround from the baseline of –$2M.).

In 1995, the plant was operating much closer to performance excellence as compared to its first full year. The five-year transition was more of a business culture transformation than a change effort. Personally, I left the organization on September 15, 1995, to start a new adventure in the consulting business with first-hand experience in the areas of leadership, strategy and organization performance. During the 90's, the primary goals for what ultimately became Affordability metastasized.

AFFORDABILITY GOALS

Operational goals can be defined in terms of quantitative and qualitative. Quantitative goals can be measured and monitored in terms of time, quality and cost. Qualitative goals are more human-type metrics and measures based on opinions, attitudes, sentiments and feelings. Combining select quantitative and qualitative goals the organization's systems and processes can be used to observe, adjust and improve operational performance. The NCR Retail Systems, E&M Atlanta facility example, established a plant performance score board based on the qualitative and quantitative measures of Customer Satisfaction (25%), Associate Satisfaction (25%), Ontime Delivery (20%), Quality (20%) and Profitability (10%) displayed operational performance and rewarded employees according to performance-to-goal (Note: 50% Qualitative, 50% Quantitative.).

Since Affordability delivers value with speed and quality at a price the Customer is willing to pay, quantitative goals can be established in terms of speed, perfection and savings. Goals of this type are more objective and can be measured using tools and methods for determining specific quantities, magnitudes and amounts. Goals of this kind are more tangible. Often, the marketplace or industry has an established timeframe for the order-to-cash cycle, a level of product or service accessibility, product perfection, errorless service and commercial worth. The value one pays for assumes time, quality and cost. If a baseline can be determined, goals can be determined to establish a target for increasing performance. The ultimate goal for Affordability can be simply stated as faster, better, more Affordable products and services that delight customers, satisfy employees and gratify suppliers.

Qualitative goals can be established based on human factors. Customers, employees, suppliers and partners can all be measured by satisfaction and delight. Goals of this type are more subjective based on opinion and perception. Goals of this kind are more intangible. It is also the case that the marketplace or industry has a recognized performance point often published in market research fact and data reports. Unless the product or service is truly unique, both quantitative and qualitative goals are available for adoption and implementation.

At the enterprise level, a balanced scorecard for Affordability can be instituted for monitoring Customer, Business, Process and People performance. Taken from the "Balanced Scorecard" from Kaplin and Norton's research and publications in the 1980's and 1990's, Affordability's Balanced Scorecard permits any establishment the fortune of measuring, monitoring and managing the four critical elements of the firm.

Always keep in mind, Goals can be set at any level. Various levels of goal setting can occur from a Strategic level of the Enterprise, to an Operational or Systems level for a facility or site or process, to a Tactical level for procedures, tools or methods, to even at a personal level for an individual. For example, a strategic goal may be an objective for a plant or site, while that objective may serve as a plant goal. An objective of 100% on time delivery may be a plant objective yet serve as a system goal for the delivery of value to a customer. A personal growth objective may be 40 hours of annual training and development yet serve as an overall organization goal for the 'People' dimension of a scorecard.

An organization's goal may be translated to operational objectives and further translated for tactical activities and then individual actions. An organizational goal may be held at the operational, tactical and individual level, such as the "People Development Goal" of 40 hours of training and development annually. Goals should provide direction, alignment, motivation, communication and execution for an organization. Goals are a future perspective of performance. Goals should take into consideration the current state or baseline, before being set and put in place for the organization to pursue. Goals should be S.M.A.R.T. (i.e., Specific, Measurable, Achievable, Relevant, Time-Bound). Finally, the amount, or 'how much', for each goal should take into consideration a true knowledge of the possibility of achievement. The story of the "Impossible Goal" presented above, had a possibility factor to it (i.e., if 50 trucks a day could be reached, the impossible would be possible). The tale of the NCR plant had the dimension of plausibility built in.

YOUR PATHWAY AND YOUR JOURNEY

With an understanding of the task at hand, an overarching goal or set of goals should be somewhat known when starting an organization. When I started my business, my short-term goals were centered around satisfied customers, successful projects, good revenue, low overhead and development of business tools that were repeatable. My long-term goals focused on client base growth, alliances and partnerships, increased market footprint, and establishment of a trusted, competent, high value consulting firm. It is typical in the first year of the consulting business that there is a 90% drop-out rate, and in the second year, 90% of the 10% left is eliminated. By the end of the third year, almost every one of the independent start-ups fail. Most of my first three years were carried out by the three projects my initial customer provided. After my first several years, my clients convinced me to personally deliver all services; hence, my goal to build a consulting firm was changed to developing and maintaining a "hired gun" approach where I would personally deliver all services as a gun for hire. Over time, the singular or individual approach evolved into a flexible robust approach using colleagues and partners to supplement additional needs and requirements when necessary.

If you're involved with an existing enterprise, organization or system, I suggest an 'environmental scan' in addition to your understanding to create a perspective of the current state conditions and develop a future state outlook. If you're seeking organization goals begin with a balanced scorecard. If you're establishing operational goals, use focal points that apply directly to the operation at hand. For tactical goals, focus on system and process performance metrics and measures. For individual goals, center on strengths, growth and personal purpose. For each tier of existence (i.e., Strategic, Operational, Tactical, Individual) goals can be established for setting direction, aligning resources, motivating people, communicating 'the message', and executing a plan. Later in this book, there are chapters explaining how to achieve linkages of goals, to objectives, to targets and to aims.

From the understanding you've gained, set one to five goals that will serve you on the path forward. Always keep in mind, goals of life may change. The terrorist attack of 9-11-2001 changed my direction for three years, until I started to get back on track toward the end of 2004. Although its easy to say, yet difficult to do, "don't hang on to a goal too long." My 9-11 consulting goals changed to operating a Coffee House in 2002, yet returned in 2004, and put me back on my path. If the goal is right, it will persevere, return and succeed.

When a goal succeeds, or set of goals succeed, leverage the accomplishments to continue, or consider an alternative path. My first career, Academia, lasted 8.5 years with successful goals and I joined NCR. My Corporate career lasted 10.5 years with goals accomplished and I created my own business. My business has lasted more than 30 years and I am currently considering my next "career." Every time, take the goals you achieved and leverage them for your next step in the journey.

Finally, I encourage you to take the risk, set goals according to your understanding of the situation at hand, and put the effort in motion to move forward. I cannot guarantee success, but I can guarantee personal growth whether you succeed or fail. However, I do guarantee one thing, there will be failures along the way, but like I always taught the kids of little league teams I coached, "When We Win, We Celebrate! When We Lose, We Learn!" Set Your Goals, Celebrate and Learn!

3 Leadership

> You manage things, you lead people. People must be led.
>
> **– Grace Hopper**

> Leadership is about setting a direction. It's about creating a vision, empowering and inspiring people to want to achieve the vision and enabling them to do so with energy and speed through an effective strategy. In its most basic sense, leadership is about mobilizing a group of people to jump into a better future.
>
> **– Dr. John P Kotter**

LEADERSHIP FOR AFFORDABILITY

On Monday, May 14, 2012, I was having lunch with Dr. John Kotter at a Legal Seafood Restaurant that used to be near Dr. Kotter's office in Cambridge Massachusetts. My purpose for the visit was to get his signature in his book "Buy In," which eventually became a gift for my son upon his graduation from Oregon State University. During our discussion, he stated a profound fact that still sticks in my mind today, "We've done a thorough job in this country teaching management and developing managers, but we haven't done a good job teaching leadership and developing leaders." Affordability 2.0 defines leadership using several attributes starting with three factors identified by Dr. Kotter in the book, *A Force for Change*, and detailed in his book *What Leaders Really Do.*

Within the book *A Force for Change*, in Chapter 2, Dr. Kotter features a past business acquaintance and former "colleague," NCR Vice President, Mr. Jim Adamson. Jim was a superior of mine whom I encountered while working in NCR's Corporate Headquarters, in Executive Vice President Darrell Clark's Banking Systems Division. Jim was a hard charging, intense, energetic Scot in charge of NCR's Banking Systems development and production in Dundee and Dunfermline Scotland. He was highly customer focused to a 'fault', and he constantly emphasized the importance of production line efficiency, quality and accountability. At one point, he placed his desk in the middle of the factory floor and welcomed anyone and everyone to personally provide input as to improvement and performance of the production line. I vividly remember a time, when we were together in Scottsdale Arizona at NCR's annual Strategic Planning Retreat, while riding horses in the desert as part of a recreation venue, I was fortunate enough to discuss some of his philosophy and learn how he transformed the culture and organizations in Dundee and Dunfermline Scotland. He had a proactive approach in which he was intimately involved with the Customers, the workforce and the systems creating and delivering Value. After I read Chapter 2 of Dr. Kotter's book *A Force for Change*, I was able to understand why he chose to feature Jim as an example of great leadership. The three actions of "Setting Direction,"

DOI: 10.1201/9781003772934-3

"Aligning Resources" and "Motivating People" were all attributes of Jim's style of leadership. And, to add, he was outstanding at "Communicating the Message" necessary to necessary to enlighten and inform the individuals working in the organization. More than anything, he was great at "Executing the Plan." With my exposure to Ret. Rear Admiral Grace Hopper, Dr. John Kotter, Jim Adamson and Dr. Demming in the 1980's and 1990's, my definition of leadership was drafted, defined, refined and evolved to what I know today.

Based on the reality, "it's not what you say, but what you do that counts," are the underpinnings that supports defining leadership in terms of actions and behaviors. Leaders, through their actions and behaviors, project their intent and shape the perception for others in the organization. For some great leaders, it's quite natural. For most leaders, there are gaps that need to be developed for optimum effectiveness. Most people in an organization form their perception of the leader, the leader's intent and the leader's expectations by what they see and observe rather than what they hear from 'leader rhetoric' and 'speeches'. Clearly stated in an article from Forbes ("The Perception Gap: Why Great Leaders Fail to Influence," by Anna Barnhill, Jul 22, 2025, 09:30am EDT), "The only thing that counts is leadership effectiveness as perceived by stakeholders. Not your intentions. Not your effort. Not even your actual improvements—only what stakeholders experience and notice."

GREAT LEADERSHIP

The best example of Affordability Leadership that I have to offer was demonstrated during a system performance improvement initiative at Northrop Grumman that began during the first week of April 2007. During the first few months of 2007, a Lean effort was designed and planned to focus on improving the processes and system that built 60% of the F/A-18 Super Hornet fighter jet. The VP responsible for Super Hornet programs was George Vardoulakis. Not only was he personally involved in the planning of the program, but he was intimately involved in the rollout, implementation and the ongoing oversight. The Super Hornet assembly plant in El Segundo, California had a sordid history of implementing Lean practices beginning in the year 2000. Initially, in 2000, the effort didn't 'stick' with the culture. Two more attempts at implementation failed in 2002 and 2004. With his new assignment in 2007, VP George Vardoulakis put a Lean initiative in place to increase plant effectiveness and efficiency. At the beginning of the year set the direction, aligned the resources and motivated the people with the rollout of a new program that took the effort beyond Lean into a greater level of performance which was ultimately, and affectionally, titled: "Affordability."

The goals of the program emerged from a sense of urgency, established by the "The Customers" (i.e., US Navy and US Marine Corps) with the need to replace older Hornet models (i.e., Models A, B, C and D) with new 'E', 'F' and eventually 'G' models. War efforts in Afghanistan and Iraq were placing a demand requirement on the military to replace older Hornets with the newer models at a rate of more than 60 per year. In 2006, only 42 Super Hornets were produced, and the demand for more than 60 seemed out of sight. Increased quantity, improved quality and decreased cost were the primary requirements for the program. Such wants, wishes and needs were much greater than a simple Lean change project could deliver. Since the Super Hornet product was contractually a Boeing aircraft, with 60% of the statement of work being Northrop Grumman's responsibility, a cultural transformation was required to achieve the customers' goals. This transformation endeavor was eventually named "Affordability."

With a few thousand people employed by the organization, Mr. Vardoulakis (AKA George) knew that he had to influence everyone for the transformation to be successful. With a good understanding of the potential problems and challenges, a familiarity of resources and tools available, and the necessary objectives that need to be met, George led, and approved, a solution design that impacted every area of the establishment. The design center of the effort was configured to use teams of six to focus on at least 50 process areas for increasing speed of production, improving quality of output and lowering cost. The Value Delivery areas, or "Value Stream" was the primary concentration of the effort. During the first week of April, three teams of six gathered for five days, each with a defined challenge and a problem to solve. George Vardoulakis personally participated as a member of the Sub-Assembly Team in that group. During the first week session, my project partner, Dr. Elizabeth Cudney and I exposed the group to the basics of Lean production, Lean thinking, Lean tools, Lean methods and Lean materials that would serve the teams for solution creation. That first week ended with a draft solution for each team to begin their implementation. A month later we came back for a second week of more advanced knowledge transfer and a status check on each team's project. This pattern repeated for each of the pair of months for the rest of the year (June/July, August/September, October/November). In December 2007, the results were reviewed and the plan for 2008 was developed. The first project completed was the solution for Sub-Assembly Team in which George Vadoulakos was a member. Throughout 2007, George continued to "Communicate the Message," keep the organization progressing toward direction and vision, and kept the resources of the organization aligned for achievement of the purpose, goals and objectives. During the end of each two-week team session, George would participate in the project report read-outs of project status and results. At the end of the first year, 12 solutions were realized and 72 people were directly involved in solving problems and getting the plant to go faster, make better and reduce cost. Affordability had gained momentum and popularity throughout the plant.

We entered 2008 with a rhythm that carried impetus and impact. Throughout the year and the next year, and most of the final year, the program continued. Each improvement event was concluded with a celebration, recognition of the accomplishments, and often, with participation of program officers from the Navy and the USMC. Customers were often involved in direct observation and recognition of implementations and outcomes of team solutions. During this second year, the program extended its reach to at least 60 of the major component suppliers for the purpose of getting everyone in the enterprise on board with the Affordability Program. A supplier conference was called, that was kicked off with the delivery of a message from Mr. Vardoulakis, with participation of the Navy and USMC representatives, and a video that was displayed and distributed to the participants which illustrated and visually presented the vision, goal and expected outcomes of the program.

Throughout 2009, and into 2010, the Affordability effort continued resulting in dozens of team improvement implementations, involving hundreds of individuals directly, and a multitude of individual contributions that affected the entire organization via support and participation. Everyone eventually participated in Affordability causing a

OFFICE OF THE UNDER SECRETARY OF DEFENSE
3000 DEFENSE PENTAGON
WASHINGTON, DC 20301-3000

SEP 14 2010

ACQUISITION,
TECHNOLOGY
AND LOGISTICS

MEMORANDUM FOR ACQUISITION PROFESSIONALS

SUBJECT: Better Buying Power: Guidance for Obtaining Greater Efficiency and Productivity in Defense Spending

FIGURE 3.1 Under secretary of defense letter to all acquisition personnel.

Page 7 Paragraph 3

The Navy, for example, recently concluded negotiations for a multi-year procurement of 124 F/A-18 strike fighter and E/A-18 electronic attack aircraft, which will yield over $600 million (greater than 10 percent) savings to the Department and the taxpayer. The F-18 program was able to drive down cost for each lot of aircraft procured in the framework of a fixed-price incentive contract that meets the Department's objectives for realistic costs, reasonable profit, a 50/50 shareline, and a 120 percent ceiling.

FIGURE 3.2 The value of the affordability program.

cultural transformation. Without leadership, this would not have happened (Figures 3.1 and 3.2). The outcome was best summed up in a memorandum distributed by the Under Secretary of Defense at the time, Ashton B Carter, September 14, 2010.

Here's what three years of leadership achieves. Speed of production increased by 59.5%. Quality improved by reducing defects per million opportunities (DPMO) by several thousand. Cost reductions rolled through to reduce the base price by $5.1M. Demand for the product also increased, and the customer base grew from only one customer (US) to at least five. It's worthy to note that the F/A-18 program was supposed to end in 2020, but it is still in use today.

Here are the results;

Affordability: F/A-18 Super Hornet Models E, F, G

	2006 Baseline	2010 Outcome
Production Capacity	42	67
Quality	4σ–5σ	6σ+
Price	$55M	$49.9M (–$5.1M cost)
Market	U.S. only	U.S.
		Australia
		Japan
		Switzerland
		Others

Without the leadership of George Vardoulakis, this would not have been accomplished. A key feature of this program is the fact that this effort destroyed the typical governmental cost-plus method of accounting and used the new paradigm of Affordability to achieve a Faster, Better, more Affordable method of production. It's worth noting that although the government realized this profound approach to acquisition, and it was communicated throughout all acquisition personnel, the practice of Affordability was not adopted as a best practice. Only recently, has the commander of the armed forces, the president, put any emphasis on Affordability. Once again, Affordability efforts, have proven that there must be a goal and vision in place, and a leader committed to pursuing Affordability for it to be realized.

AFFORDABILITY LEADERSHIP

Affordability Leadership is defined through the actions and behaviors of great leaders. My first memorable exposure to leadership came in the early 1960's growing up in south Florida. Other than my Mom and Dad, my teachers, my baseball coaches, my Cub Scout pack leaders and of course, the TV shows featuring western good guys and American combat soldiers formed my original early understanding of what leadership looked like. One of my most unforgettable experiences with leadership occurred on a Cub Scout field trip to the edge of the Everglades near Carol City where the US Army established a missile base during the Cuban Missile Crisis. The father of a member of Our Cub Scout Troop was stationed on that base, and he was able to arrange for us to visit the site. We at lunch in the mess tent, we were escorted around the camp, and we were even taken into a bunker control center where missile launch equipment was controlling the missiles aimed at Cuba. What fascinated me more than anything were the actions and behaviors of the base officers. I don't remember exactly what they said, but I do remember their presence, their actions and their behaviors. Without knowing it, that experience, and the experience of others in my formative years, shaped the foundation for my understanding of leadership.

Fast forward to June 1969. As a teenager, I was able to witness a leadership commitment that was made several years before by President Kennedy to go to the moon (i.e., September 12, 1962). After that time, a great deal of effort and funding was provided to the National Aeronautics and Space Administration (NASA) that focused an entire country on that goal. Through the 70's and into the 80's, my models of leadership remained to be my parents, my teachers (and professors), my coaches and the management of the organizations I was employed by after I graduated college in 1976. During the early 1980's, as I pursued my Master's Degree, I had a brief encounter during an event in 1982 that changed my understanding and definition of leadership for life. That summer, I was scheduled to speak at a computer conference at the North Charleston Conference Center Charleston South Carolina and present the research and work I had been doing with micro-computers at all levels of learning from pre-school to post-PhD. My academic concentration was in math and computer science, with an emphasis on how humans learn with computer systems and what was required to transfer knowledge to children, adolescents, young adults and even very seasoned individuals, some with PhDs at the University of South Carolina. As I was preparing to give my presentation, the keynote speaker, Ret. Rear Admiral

Grace Hopper was speaking on the history of computing and the future of computing. As I nervously waited for my turn on stage, I heard her say one of her famous quotes, "Remember this … 'You manage things, and you lead people.' (she added) People must be led'." From that moment forward, my intrigue and research on leadership began.

When I left academia, and joined corporate america, working for NCR in 1985, my definition of leadership continued to mature. As part of the General Computer Systems Division, in Columbia South Carolina, I broadened my comprehension of leadership through encounters with numerous successful managers and executives. After a few promotions, and a transfer where I was "voluntold" (i.e., told to volunteer) to go to NCR Headquarters in Dayton, Ohio to work on Strategic Planning efforts, my understanding was considerably broadened to include exposure of leadership (good and bad) inside the company and outside through customer experiences. That was the timeframe I encountered Jim Adamson in Scottsdale Arizona, during a Strategic Planning Retreat. Later that decade, I was once again 'voluntold' to join a new effort to institute a Just-In-Time assembly plant in Duluth Georgia, where, through the connection of AT&T Bells Labs in 1991, I was able to gain exposure to the philosophies of; John Kotter, Peter Drucker, Thomas Edison, W. Edwards Deming, Jack Welch, Teresa Amabile, Robert Kaplan, Michael Porter, James Kouzes, Barry Posner and others.

By the time I left NCR and started my own business on September 15, 1995, I had enough experience and exposure to formalize my definition and description of leadership. Two events happened in 1995 that further increased my competency and boosted my aptitude for leadership. I was invited in to participate in Dr. Teresa Amabile's Harvard Business School Research Study focusing on motivation in the workplace (AKA The Events and Motivation Study or T.E.A.M. Study) and subsequently, also exposed to the work of Dr. John Kotter who published his renowned book, "Leading Change" in 1996. By the end of the 90's, my definition of leadership hung on ten points divided into actions and behaviors.

In 1999, when Dr. Kotter published the book *What Leaders Really Do*, the lightbulb went on. In Chapter 2 of that book, he writes about the difference between management and leadership. He goes on to clarify the skills leaders need. First of all, "Setting a Direction." Second, "Aligning People." Lastly, "Motivating People. He closes the chapter with a discussion of creating a Leadership Culture. These three attributes and actions are critical for leaders to institute change and transform organizations for the better. Since management is about things, and leadership is about people, and the movements and mannerisms project the leader traits of an individual, leadership can be defined through actions and behaviors.

Unlike other models of leadership, Affordability Leadership uses an actions-and-behaviors based instrument that is based on critical aspects defined through the observation and analysis of great leaders. There are five actions and five behaviors that serve as variables to assess and predict the potential and probability for success of any leader. The five actions include Setting Direction, Aligning Resources, Motivating People, Communicating the Message, Executing the Plan. The five behaviors consist of Modeling the Way, Inspiring a Shared Vision, Challenging the Process, Enabling Others to Act, Encouraging the Heart. When combining all ten, a profile of the leader is created and it can be used for selection, improvement, appointment or even change and dismissal.

The five Actions are described as follows:

1. Setting Direction: The ability to determine, define and describe the path forward for a group or organization. Distinction and clarity is of the utmost importance. Of all the great historical accomplishments observed, direction sets the course and defines the route. Today, numerous organizations have a defined purpose (with principles and values), vision, mission, goals and objectives that are meant to convey direction. In this instance, its not a written document, but rather a leader's actions by word and deed that express a collective aim. When President Kennedy said, "We choose to go to the moon …" the statement itself, and the next several years of actions demonstrated strong leadership direction.
2. Aligning Resources: First and foremost, it's the alignment of people. The right people, in the right places, with the right talents, at the right time, for the right reasons. Along with the human factor, the systems and processes of the organization require the alignment of technology, machines, tools, materials, methods and money. For Affordability, alignment with the Value Stream, the Value Stream Support Resources, the Suppliers and even the Customers require alignment for success.
3. Motivating People: Whether it be sense of urgency, pursuit of a great cause, need for survival or just recognition of a job well done, motivation is a critical component of successful organizations. Motivation is a guiding force of human behavior. Some inferior organizations have become successful over their superior competition by motivation alone (Suggestion: watch the movie "Miracle on Ice." I watched the event live, and from that point forward, I was convinced of the power of motivation). Great leaders motivate people to do things they might normally not do on their own.
4. Communicating the Message: Over the past 30 years, and even the 20 years before, the most often cited shortcoming and weakness I've heard expressed is, "Communication." It's not necessarily the amount of communication, nor the lack of communication, but often the content, truth and integrity of the communication. At times, there is too much communication which causes apathy, confusion and even chaos. At other times, there's too little communication that can cause controversy, conspiracy and doubt. In Affordability, the message containing Value and the Value Stream must be clear, concise and communicated in multiple forms (e.g., spoken by the Leader, spoken by the management, written and distributed in various forms including emails, newsletters, board notices, training and even through hands on experience in the work and processes of every individual). "The Message" may contains a new direction, an organization transformation, a major milestone accomplishment, or any other critical success. When working with Bell Labs Organization Effectiveness Center in the 90's I was part of a study group that came up with "The five Times Rule": If a message is the utmost importance, it must be communicated at least five times, in five different

ways, using five different types of media, by five different leaders, in five different settings (which mathematically comes to 625 times). Great leaders have a natural (or maybe even developed) ability to communicate the message in a manner in which the message is understood by everyone.

5. Executing the Plan: It's not necessarily the plan, but it is the planning, and the execution of the plan, with ongoing planning that is paramount to success. Leaders who have the ability and capability to see a plan fully implemented, prove to be the most successful leaders over time. This action ties back to the other four. For a plan to be successfully executed, direction must be set, resources must be aligned, people must be motivated and the message (the plan) must be clearly communicated. In addition to the other actions, great leaders are often intimately involved with carrying out the plan to the extent of direct participation.

The five behaviors are described as follows (Specific credit must be given to Professors Kouzes and Posner and the publication "Leadership Challenge"):

1. Model the Way. All too often, leaders give orders or provide instruction thinking that people will automatically follow their commands. Some will, some won't and some will wait and see. If the leader has always been true to his or her word, and has, over time, demonstrated integrity through action, the people in the organization are more likely to follow any directive the leader provides. There's an old phrase that applies here, "Do what you say, say what you do, but mostly, you must do." To 'Model the Way' refers to being an example of what you expect through behavior. Leaders who do this create advocates in the ranks and develop cooperation and collaboration toward the goal, vision, mission and objectives. People will follow a leader who demonstrates a willingness to carry out the expectations he or she has of others.
2. Inspire a Shared Vision. This behavior goes beyond the physical and intellectual, realm and reaches into the emotional and spiritual domain of individuals. It directs leaders to motivate their people to go forward toward a mutually understood direction. A goal common to all, believed to be achievable by all, aligning with a mutual purpose is that shared vision. This particular behavior ties back to the first three actions stated earlier.
3. Challenge the Process. Historically and traditionally managers were directed to challenge the people. This behavior is more collaborative in nature since it focus the people on the process and the work. It questions the validity and legitimacy of the process, which permits the leader to apply the people for improvement. Old style management was too often encouraged to continue replace people in effort to 'right things', when all along the process was broke. Dr. Deming used to say, "The people aren't broke, the process is broke. Use the people to fix the process."
4. Enable others to Act. This happens when the leader focuses on the development of the people, encourages and supports the people to accomplish tasks and take on challenges, and recognizes achievements and efforts.

Enabling often requires taking away long-standing restrictions that created roadblocks and barriers prohibiting success. It is sometimes said, the organization itself is sometimes the worst competition.

5. Encourage the Heart. Every organization has heart and soul, emotion and spirit. Some are energetic while others are lackluster. A leader who energizes their establishment will benefit by the energy of all of the others. We often hire people for their 'hands' or their 'heads', their capabilities and talents, but the best of the high performers have their hearts and spirit in the game.

The 10 Points of Leadership described above can be used as an assessment tool on any and every leader in the organization. An assessment can be performed by an individual or group that has a common understanding of the degree to which person performs. Applying a 1-to-10 scale, a score for each characteristic can be determined and an aggregate score can be compiled as well as a focus on areas for improvement.

If the scale was anchored with 'descriptors' for evaluating a leader's performance, an assessment can be implemented for evaluating each leader:

Very Poor			Poor		Average		Very Good		Excellent
1	2	3	4	5	6	7	8	9	10

Even for good managers in an organization, scores of 9 or 10 are usually not common. When compiling the aggregate score, add up all the average scores for each of the ten categories. If done with honesty and integrity, scores in the 90's are rare. Good leaders fall into the 80's, average leaders in the 50's and poor leaders below 40. Keep in mind, this approach is not meant for "public broadcasting event." It's a one-on-one situation between a leader and his or her superior. Scores are to be kept confidential.

I've used this instrument with clients, CEOs, Executives and partners. I've even used it to evaluate the US Presidents I've been familiar with since the 1960's. After testing, piloting and using the 10 Point Assessment since 2000, I'm convinced it's quite accurate and provides a predictable outcome for leading Affordability efforts, as well as supplying a leader with some input as to how to improve.

YOUR PATHWAY AND YOUR JOURNEY

Assess yourself using the Affordability Leadership instrument. Evaluate yourself on all ten points. If there's a leadership team, assess each individual and hold a team self-assessment session to get a comprehensive understanding of the current state of leadership as well as the opportunities for improvement. I've used it on myself and I've used it with clients for transformation, change and improvement projects. On one occasion, a leader, my customer, had me perform an assessment of him using his reports, and select employees of the organization. The outcome was quite valuable in providing insight for the leader's improvement, the leadership team's improvement and the organization's improvement.

Adopt the ten points of leadership as the model for you and/or your organization. Keep in mind, people judge leadership from the actions and behaviors of leaders, and are further convinced of a leader's profile from outcomes and results of those actions and behaviors. From the adage "success breeds success," as more success is observed by the people, the more success actions and behaviors are emulated by the people.

Set Direction. Set a direction for your team or your people so that they are aware of where you intend on going. Whether it be a stated vision or mission, or stated purpose, or even principles and values accompanied by the path moving forward. People want to know what's next, what to expect and where they should put their efforts. If a direction is not established, people will establish their own path forward. In order to focus everyone's energy and efforts, direction is necessary.

Align Resources. First of all, align the people of the organization with direction, purpose, vision and mission. After the people are aligned, be sure and align all supporting resources with the people of the organization. This refers to systems, processes, machines, tools, resources, materials and even money. Alignment also applies downstream to customers and upstream to suppliers and partners. Ultimately, the alignment of the entire enterprise is the overarching goal. This includes scorecards for communicating metrics and measures for; Customers, Business, Processes, People.

Motivate People. Although money is a motivator, money is a short-term motivator as long as the money lasts (Note: If money is always applied as the motivator, eventually over time, motivation money becomes a dependency and fails to be a motivator.). When recognition is applied as a motivator, keep in mind that you, the leader, is a good source for motivation, as is other leaders, managers, coworkers and even customers. Awards and rewards (including money) that is provided upon achievement or accomplishment motivates, and so does a genuine publicly presented 'job well done'. Other motivation methods include: promotions, personal development, new opportunities, time off, team celebrations, organization commemorations.

Communicate the Message. Personally be involved in the communication of critical information. Use the leadership team, bulletins, bulletin boards, email, get-togethers, team gatherings, group meetings, etc. When a critical message is crafted, define methods to actualize and emphasize the message. For example, if customer satisfaction is an important focal point, include customer testimonials to emphasize specific points to clarify. Also for example, if quality is an important subject, publicize and broadcast any quick wins gained from improvement efforts. Bottom Line: If it's important, communicate it concisely, clearly and consistently.

Execute the Plan. Although the plan is likely to change over time, ensure it's success is realized. Continued stops and starts, or continuous re-starts, sets a bad precedent, so the continuum of the pursuit of completion for a significant project is vital (Warning: Don't start a big project or serious effort if you're not going to personally see it through.). Great leaders display the ability to successfully see through the completion of major projects (Note: Weak leaders seem to not be able to get things done.).

Model the Way. "Say what you'll do and do what you say." Behavior of the leader, especially in the area of integrity, builds trust and confidence in the group or team or organization being led. The people watch the leader in an attempt to see if the leader displays the behavior he or she states will, or should take place (e.g., If a leader commands, "Take the initiative and correct the problems you see." The people will watch to see if (s)he goes after the organization problems that are clear to everyone, and solves enterprise, systems, or process issues.). Do not convey, "Do as I say and not as I do." (e.g., I remember back in the days of COVID, when some political leaders telling others to "wear a mask and isolate yourself," while all along they were recorded not wearing masks and intermingling in public with others.). Model the Way is meant to help a leader build integrity, trust, honesty, reliability and confidence in the people.

Inspire a Shared Vision. Tied to the actions of 'set direction', 'align the resources' and 'motivate people', 'Inspire a Shared Vision' encourages a behavior that gets everybody 'on the same page' by promoting a collaborative vision that the people can 'get onboard' and pursue a common and cooperative goal. An age-old adage clearly conveys the importance of this behavior; "Without a Vision the People will perish!" Without a vision, or mission or direction, the organization will likely perish. Here are some recent companies I've encountered that have perished; Sears, Rite Aid, Woolworth's, A&P, American Motors, DeLorean, Sun Microsystems (Note: All suffered from lack of vision, mission, direction.).

Challenge the Process. More often, the systems and processes of an enterprise have issues and problems, independent of the people and the work being performed. As Dr. Deming often put is, "The people aren't the problem, the process is the problem. Use to people to fix the problems in the process." By challenging the process, a leader is afforded the opportunity to focus the people on the problem, and align the proper resources to fix the problem at hand. The more process problems fixed, the more Affordability is achieved.

Enable Others to Act. Whether you institute training and development, or invest in process and quality improvement efforts, or institute a standard problem-solving process for use throughout the organization, methods, means and tools that enable people to work better have proven to motivate and inspire a workforce. Personal involvement of leaders and managers within the labor force has proven to be beneficial over time. Providing freedom for employees to solve their own work problems and improve their own work environment, processes, methods and means have been researched and recognized to be advantageous to every organization.

Encourage the Heart. When people only work with their hands and their heads, average results are often the case. However, when people have their hearts and souls involved in their work performance excellence is usually the outcome. When a leader is engaged, and the people are inspired, the conclusion is success.

Use the ten points of Affordability Leadership for yourself, your leadership team and your people. You might even encourage your suppliers and customers to adopt the philosophy and institute it within their organizations. Many models of leadership have been discovered and used by many organizations; Affordability Leadership is designed for any type of organization and optimized for the pursuit of Affordability.

4 Strategy

> Strategy without tactics is the slowest route to victory. Tactics without strategy is the noise before defeat.
>
> **– Sun Tzu**

STRATEGY FOR AFFORDABILITY

As most people understand, a strategy is a method for achieving a goal. People often assume that executing a strategy requires a great amount of time. Most often, strategy execution does require "a great amount of time" that could amount to five years, ten years, or more. I've even been involved in successful strategic endeavors that have taken as little as three years up to as much as ten years. The timeline for the US's involvement in WWII started on December 7, 1941 (i.e., the bombing of Pearl Harbor) and ended on September 2, 1945 (i.e., the Japanese surrender) amounting to four years and nine months accomplishing the two goals of defeating both Germany and Japan. For a strategy to be successful, it requires strong leadership, a clear strategy, a strategic plan (or, strategic plans and planning) and a collaborative goal or set of goals (i.e., two or three).

The best successful strategies I've witnessed were carried out by strong leadership with a good plan. Often, the plan, over time, evolves to meet and mitigate challenges that arise as conditions change. Good plans are often architected using a layered approach, where typically, three or four layers are used to describe the details of actions and activities from the major goal stratum through operations and tactics to the team and individual tier for execution. The overarching layer of a strategic plan covers the entire strategy or organizational construct and represents the "Strategic" or "Enterprise" level. Strategic objectives are applied at the operational level, which in turn are measured and monitored at the tactical level, that direct the tasks at the team and individual level. In essence, there's the linkage of goal(s) from strategy to operations, that is also linked through tactics to the teams and people carrying out the mandatory tasks.

Strategies can take on different forms using different formats within various realms (e.g., Organizational, Operational, Market, Product, People). The focus could range across a variety of diverse scopes: Business Unit, Program Level, Enterprise, Corporation, Functional, Operating, Financial, Target Market, Competitive Product, etc. Shortcomings of some strategies happen when some of the five basic building blocks for success are weak or missing (i.e., 1. Vision; 2. Leadership; 3. The People; 4. Resources, Methods and Tools; 5. Design and Plan).

Strategy can be driven from a sense of urgency or executed from well-crafted strategic plan(s) or both as it was in WWII. Strategic Plans are meant to convey direction, alignment, motivation, communication and execution. They are the purpose,

DOI: 10.1201/9781003772934-4

reason and mission for the organization to pursue. Strategic Plans, once written, are highly likely to change over time as circumstances, conditions and situations shift and transform. The ongoing planning is of utmost importance, although Dwight D. Eisenhower was once quoted to say, "Plans are useless, but planning is everything." ... however, the plan's purpose is to communicate the strategy, goal(s), objectives, tactics, and actions, as well as what the leader and leadership team expects from the people at every level (i.e., Strategic, Operational, Tactical, Team/Individual).

The primary elements for the Strategy for Affordability involves Value, Customer, Cost and the AIM addressing all three major components simultaneously to achieve profitability and prosperity. For any Strategy for Affordability to be successful, the five variables of the success formula (i.e., Vision + Leadership + People + Resources and Tools + Design and Plan) should be well defined in the design and plan for execution. Once all of the components are well understood and defined, and the design is known, the development of the plan can begin.

STRATEGIC PLAN

A strategic plan for an organization is meant to communicate mission, vision, direction, purpose and goal(s) for the organization (Note: Also, Principles, Core Values, Sense of Urgency). The contents of a strategic plan should convey what is being done, how it will be accomplished, where everything will take place, who will be involved and when each step and stage should be executed.

Clear Leader Direction is a necessary element of effective strategic plans. Whether its vision, mission, objectives or course, alignment, motivation or purpose, urgency, expectations or "all of the above," the leader must provide the organization with a stated direction within the plan. The leader and the leadership team must be intimately familiar with, and engaged in, the development and deployment of the plan. The more that leaders involve themselves, and the more integrated they are into the action and activities, the better the outcome. The People of the organization are energized and motivated with leadership involvement

Enterprise-wide Systems and Processes, along with the Value the Organization provides, should be clear, distinct and unmistakenly unambiguous. Value should be at the core of the execution of the plan. People will "fight" for Value if they believe in it. I've seen it on the battlefield, in hospitals, in successful businesses and in non-profit establishments. Also true, when the "fight is on" people sometimes ask themselves, 'is this worth it?'. The strategy, strategic plan and execution of the plan should be 'worth' the effort. The Value of what's being done and the Value of the undertaking should be crystal clear to everyone.

When "the game is on," the people need to know "the score" and if they're winning or losing. Beyond just scorecards and scoreboards, performance goals, objectives, baselines and targets should all be tracked, monitored, communicated and reviewed with the people. Knowing where we've been, where we are and what we're aiming at to achieve expresses specific direction, resource alignment that serves to motivate the forces. Set targets, measure success and failure, convey expectations through metrics and keep score to identify wins to celebrate and losses in which to learn and improve.

Although many organizations create and execute a strategic plan for their organization, few do a good job of creating a strategic plan for deploying their Value. Whether it be a new invention or service, or an enhancement to an existing product or business service, its most often not a common practice of strategically introducing and deploying Value.

A STRATEGIC PLANNING FRAMEWORK FOR DEPLOYING VALUE: THE VALUE DEPLOYMENT PLAN

Value within the Affordability construct is delivered by way of a product, or a service, or both. Over the past 40 years, I've observed that many organizations tend to departmentalize how a product or service will be presented and deployed to their target market and target customer base. Sales and Marketing will prepare their plans for selling and marketing. Customer Support and Technical Support will prepare their plans for serving up support. Logistics prepares their plan, while installation prepares their plan, while order fulfillment prepares their plan, and on, and on, and on. In a few specific situations, I've seen comprehensive plans created that clearly communicate a strategy for a new product or a new service and convey how the Value is to be implemented and activated. My first experience with this approach happened when I was exposed to Proctor & Gamble's product deployment methods as a Strategic Planner with NCR. The Deployment of Value Plan (VDP) is meant to be a comprehensive blueprint for how to institute and install the product or service in the marketplace and make the product or service available to all potential customers. Below is the outline that details how to prepare a plan for the Deployment of Value (i.e., The Value Deployment Plan or VDP):

- Value Proposition: Summarize the value of the product or service being offered. Provide a brief description of how customer requirements are met. Briefly describe the market and the types of customers targeted in that market. Explain how the product or service is worthwhile for the customer to purchase. The Value Proposition should explain why the product or service is worth purchasing. Describe the overall business opportunity for the product or service and include information on the competition and competitive offerings. Also, define the market size and the overall business potential of the product or service offered.
- Customers: Define and describe the targeted customers, target markets, target industries and name specific target customers if possible (e.g., US Department of War, WalMart, State of Florida, Ford, etc.). Briefly list the requirements, needs, wishes and wants that are being addressed by the product or service. Discuss why the product or service is worth purchasing. Detail as much as possible all customer profile(s) and synopsis' of customer characteristics and types. Convey market sizing and potential market share percentages.
- Competition: Identify the competition within the target market(s). Provide as much detail as possible to competitor offerings, competitor strengths

and weaknesses, competitor gaps and competitor shortcomings. Size the competition and define what portion of the competition is the target for the product or service. Describe how the product or service will beat the competition's offerings and gain market share.

- Product and/or Service Positioning, U.V.P. and U.S.P.: This section of the plan should serve as the most critical element for communicating the value and the worth of the product or service. It should serve to position the product and service in the marketplace and communicate how the product or service meets customer requirements and rivals the competition's offerings. The U.V.P. (Unique Value Proposition) clearly identifies what is exceptional and distinct about the product and service as compared to competitive products. The U.S.P. (Unique Selling Proposition) is the information and direction for selling the product or service in the marketplace against the competition. The worth and appeal for the customer should be defined in detail, and the significance and impact of the product or service in the marketplace should be positioned to combat competition.
- Product and/or Service Details: This section is self-explanatory. It should contain as much detail as possible about the product or service (Including: illustrations, diagrams, designs, fact and data sheets, etc.). Anyone reading this plan, should understand from this section everything about the functionality and features of the product or service.
- Marketing: A substantive marketing section provides the "how" a product or service will be presented to the marketplace, and the "what" needs to be planned for with tactical positioning for approaching a target market and/or a target customer.
- Sales: A fundamental sales section will specifically direct sales teams and individuals as to the "what, who, when, where and how" the product or service will be sold. The UVP and the USP should be intimate knowledge of the force deployed to sell the product or service (Note: On rare occasions, the product or service is in great demand and sold through repeat business or referrals such that a sales force is not required.).
- Support: Support can be defined in a few ways. Customer support, product or service support, and organization support for those creating, producing and delivering the UVP. The support section of the plan identifies what support is available and the types of support that apply to the product and/or service.
- Training: The transfer of knowledge and capability is training and the development of people that may be buying, using, selling, marketing, supporting and managing the product or service is capability. A comprehensive training program should accompany every product or service. A complex product or service may require more training and training methods than a simple straightforward product or service.
- Collateral/Media/Material: Substantive information needs to accompany and support the product or service and the individuals in the Value Stream that deliver and support the Value, UVP and USP in terms of documents, manuals, use instructions, fact sheets, brochures, marketing literature,

presentations, videos, etc. Every means of communicating details of a product or service should be included in this section.

- Pricing/Purchasing/Ordering: Anything related to pricing, purchasing or ordering should be covered in this section of the plan. This is the section of the plan that explains "order-to-cash" cycle of the flow of money from the customer and the product and/or service to the customer. For additional product and/or service flow information, reference the Distribution/ Logistics section of the plan.
- Distribution/Logistics: This section of the plan details the ways, methods and means the Value Stream delivers the Value to each customer. Whether a physical and tangible product, a virtual service (e.g., online banking or investing, streaming training, security monitoring, etc.), a 'material' service (e.g., drive-up food service, maintenance service, pool cleaning service, etc.), or a combination of product and service (e.g., coffee shop, steakhouse, entertainment, etc.), it must be clear how the Value will be delivered to the Customer.
- Financials: This section specifically focuses on how the "order-to-cash" cycle works financially and typically includes the budgeting and operational financial proformas. The revenue, the margins, the profits, the costs, the expenses, the spend and the comprehensive balance sheets in terms of models, forecasts, expectations and actuals are all described.
- Additional Information: This portion of the VDP is 'more or less' the appendix of the plan that contains 'any and all' additional information required to comprehensively support the Delivery of Value.

Accompanying a Value Deployment Plan should be a well-planned structure with systems that support the entire Value Stream. The Value Deployment Plan should be part of the strategy and strategic plan for the enterprise. Whether the Value being deployed is a product (e.g., automobiles, skyscrapers, healthcare implements, food, etc.) or a service (e.g., grooming, air travel, hotel/motel, military, government, etc.), it represents the worth, benefit, significance and benefit for the enterprise. The Value Deployment Plan should accompany the Enterprise's Strategic Plan and fully describe the Value being provided by the Enterprise. The alignment of the Enterprise Strategy, the Strategic Plan and the Value Deployment Plan provide direction, alignment, motivation, communication and execution throughout the Enterprise.

WHEN STRATEGY AND TACTICS COLLIDE ... A POOR APPROACH TO STRATEGY AND STRATEGIC PLANNING

Just over 200 years ago, in 1775, the Postal System of the United States of America was created about a year before the term United States of America was formerly declared in 1776. Since that time, the United States Postal Office Department, now called the United States Postal Service (and "USPS" as of 1971), operates in approximately 33,641 locations, employing more than 600,000 people, operating more than

235,00 vehicles, delivering around 127.3 billion pieces of mail (including packages on foot and by truck) to at least 164.9 million delivery points. In the 200-year period that the USPS existed, you would expect the USPS Value Stream to be well defined, tuned and operating at a high rate of Affordability. The Value of USPS is within the Value Stream of the Flow of Letters and Packages to and from citizens and organizations in the US. Before we discuss their Strategy and Strategic Plan, I'd like to unveil some facts and statistics for understanding. Here are some details and data that provide for a full understanding of the current state of this enterprise. USPS 2025 results (highlights):

- *Operating revenue was $80.5 billion, a 1.2 percent increase, due largely to growth in USPS Ground Advantage and strategic price increases*
- *Net loss decreased $542 million compared to prior year while controllable loss increased by $900 million, reflecting lower than expected revenues from our package business*
- *Transportation expense reductions of $422 million reflect continued progress of network optimization,*
- *Controllable loss, which excludes certain expenses that are not controllable by management, was $2.7 billion for the year, compared to $1.8 billion for the prior year.*
- *Net loss for the year under generally accepted accounting principles (GAAP) totaled $9.0 billion, compared to a net loss of $9.5 billion for the prior year (This decrease in net loss is attributed to our operating revenue increase of $916 million, transportation expense reductions of $422 million, and a decrease in workers' compensation expense of $1.1 billion, partially offset by increased compensation and benefits expense of $1.7 billion, and higher other operating expenses of $221 million.)*
- *"In surveying the results of the past year, the occasional appearance of financial progress—such as our profitable first quarter—is far outweighed by the reality of our significant systemic annual revenue and cost imbalance," said Postmaster General David Steiner. "To correct our financial imbalances, we must explore new revenue opportunities and public policy changes to improve our business model. Most importantly, we must operate more efficiently and compete more effectively to best perform our public service mission."*

Here is a summary of their plan for moving forward:

The Postal Service is seeking further administrative and legislative reforms to remedy outdated and unwarranted financial and regulatory burdens. These reforms include changes in retiree pension benefit funding rules for the Civil Service Retirement System (CSRS) benefits, diversification of pension assets, raising the statutory debt ceiling, and workers' compensation administration reform.

Total operating revenue was $80.5 billion for the year, an increase of $916 million, or 1.2 percent, compared to the prior year. The increase was due largely to continued growth of our *USPS Ground Advantage* shipping service as well as strategic price increases in both of our mail and shipping categories.

First-Class Mail revenue increased $370 million, or 1.5 percent, on a volume decline of 2.2 billion pieces, or 5.0 percent, compared to the prior year. Marketing Mail revenue increased $350 million, or 2.3 percent, on a volume decline of 764 million pieces, or 1.3 percent, compared to the prior year. Shipping and Packages revenue increased $315 million, or 1.0 percent, on a volume decline of 415 million pieces, or 5.7 percent, compared to the prior year.

Total operating expenses were nearly $89.8 billion for the year, an increase of $317 million, or 0.4 percent, compared to the prior year. The overall increase in operating expenses was primarily due to increases in compensation costs and other operating costs, including the incentivized voluntary early retirement offer to certain employees, partially offset by the impact of discount rates on workers' compensation costs and lower transportation costs.

"The financial results reflect the difficulties of our mandated cost structure and the continued decline in volume, offset to some degree by the Postal Service's efforts to push back against those trends by aggressively managing the costs we can control and by the judicious use of our pricing authority," said Chief Financial Officer Luke Grossmann. "We remain focused on moving toward financial sustainability through growing our *USPS Ground Advantage* package business, creating additional operational efficiencies, and developing product strategies to generate growth."

After reading the results and the intent for the near-term future, I ask, "What did you notice? What did you realize was missing? What else were you expecting?" Although the financials were quite detailed and comprehensive, there was no mention of "Customer" and no mention of "Value." In order to reach an Affordable and profitable state, the relationship between Value, Customer and Cost must be well addressed and balanced. The Aim should be Delivering the Value the Customer requires at a Price the Customer is willing to pay, and operating at an expense that proves profitable, or at least "break even."

My interest and intrigue with USPS goes back about ten years when I was serving on a USPS project operated by the IISE (Institute of Industrial and Systems Engineers). The focus of the project was to observe, study and accumulate data regarding the operations of the postal carriers of the USPS. We studied inner office operations and field operations for the purpose of gaining a better understanding of the flow from the time the mail was received the distribution center, to the time the mail was prepared for delivery, to the time the mail was delivered (and picked up at the mail stops), to the time the carrier returned to the Post Office and placed the received mail in processing. It was a cooperative effort between the USPS and the Postal Workers Union to get a good handle on the current standards and processes, and with the accumulated data, set new standards for the system and processes for Carriers. After studying more than 50 carriers myself (Note: I was one of several researchers), the data for hundreds of carriers were compiled and presented to the Carriers Union and USPS. Although I have not seen the results of the study put into action, I was able to write a summary report of all the studies I completed from a system level perspective. Given the tools and technology the carriers had to operate with, their processes and procedures were stable and predictable. However, I noted that a major issue occurred when the distribution center delivered and picked up mail (including packages). Carriers were often delayed getting out to deliver and then late in returning after their route was complete. This was exacerbated during their "heavy flow season" beginning in late October and running through the rest of the year. To put it in simple terms (in a "Demming-Like" Manner), "The Postal Carriers weren't broke, the USPS System was broke and needed to be fixed."

The strategy does not exist within the tactics of delivery. The financials alone cannot be fixed unless the system is fixed. The focus on the Value and the Customer must be the primary crucial focal point. Tactics and strategy must be aligned through operations and sound operating systems and processes. "Money, and money alone cannot be the pathway and means for fixing USPS." Until the Delivery of Value is better understood by the organization and the systems meet and exceed the Customers' requirements, the enterprise called USPS will not reach breakeven nor become independent of government subsidies. Due to long-term government intervention, USPS is likely to remain a dependent organization, requiring its primary focus to be fixed on government support and subsidies. In order to become Affordable, USPS will have to endeavor a dramatic transformation. This would require a vision, strong leadership, alignment of the people, tools and resources, and a design and plan for Affordability.

TACTICS BEFORE STRATEGY OR STRATEGY BEFORE TACTICS?

At the outset, it's an interesting query that can often be defended from both ends depending on the perspectives that arise from the current conditions. However, lazy organizations too many times look at the answer from an internal operational perspective and work upward to create a strategy. Or they begin with their current state of operations and create a strategy starting with the viewpoint of today and concentrate on improving their tactics from an inside view. In either case, it's likely to end in defeat because it doesn't include their customers' realities and wreaks of tactics without strategy. Phrases related to this topic are commonly quoted; "A Tops Down Approach," "A Bottoms Up Approach" or "An Inside Out Approach." And, as a consultant, a cliché phrase I try to avoid (although common in the consulting world), is "It All Depends." No matter what the view, the Value and Value Delivery for the Customer must be at the center of all strategic planning. If tactics are to be leveraged, strategy must be known.

A STRATEGY TO SAVE LIVES (TACTICS BEFORE STRATEGY)

The MRAP Program, mentioned earlier in this book, was off to a slow start in 2007. For the first few months, less than a few hundred trucks were fielded. After a great deal of effort, using Lean and Six Sigma methodologies and techniques, the site was beginning to meet a 50 per day goal that was required for 2008 and 2009 in order to field all 25,700 vehicles during the three-year promise made by Defense Secretary Gates. The first day to meet the goal of 50 was November 4th of 2007. By the end of

2007, more than 1,200 trucks were either delivered or on their way to protect lives. In January 2008, the rate of integration for deployment was operating at a rhythm of 50 per day for every day of production. That meant a Navy transport ship left the Port of Charleston each week (@ every five days) for Kuwait loaded with MRAPs ready for use in Iraq and Afghanistan. Each truck protected, on average, 6 lives. Protecting the lives of Warfighters was priority #1 for the US Military.

In this case example, tactics came before strategy. The vision and mission for the MRAP strategy was clear and straightforward, and our stated Purpose was: "We Protect Those Who Protect Us." The sense of urgency guided and directed all efforts to field trucks as fast as possible to protect the lives of our Warfighters. A formal Strategy and Strategic Plan was not required in 2007, because lives were at stake, it was time to act, and the time was now! The effort, in the beginning, was led by the USMC general authority, and it impacted on the Navy, the Army, the Air Force and the Special Operations forces. The operational center for vehicle integration was logistically positioned at SPAWAR Naval Base in Charleston South Carolina. At the beginning of 2008, the facility was meeting the demand rate to appropriately fulfill the overarching goal of 25,700 in three years, so the Program Manager Paul Mann assigned Col. Rob Chappell and I the task of creating a three-year Strategic Plan to serve as the communication vehicle for the duration of the Program's initial actions and activities of deployment. The window of 2008–2010 was the established time-frame for the plan. With input from all the principles of the program, involvement of more than 200 program team members and participation from numerous military personnel, the plan was published and distributed to all sites participating in the MRAP deployment. The contents of the plan were as follows:

- Cover: Full color cover with the program's official seal, title ("Strategic Plan 2008–2010 Joint Program Office Mine Resistant Ambush Protected Vehicles")
- Inside the Cover:
  - Mission: "We deliver survivable, fully capable, MRAP vehicles to our Warfighters and customers. We demand and support maximum readiness from our MRAP vehicles once delivered. We operate with speed and a sense of urgency always."
  - Purpose: "We protect those who protect us!"
  - Core Values: "We are Responsive, Adaptive, Motivated, Perseverant"
  - Vision: "We are a cohesive, Warfighter focused, rapid response JPO meeting existing and emerging needs with unparalleled tactical wheeled vehicle force protection systems that are relevant, ready and sustainable during all phases of the life cycle."
  - Table of Contents:
    - Forward
    - Program Overview
    - Program Direction
    - Leadership
    - Systems and Processes
    - External Relationships

- Forward: The Program Introduction Statement signed by: Paul D Mann (Joint Program Manager), David K Hansen (Principal Deputy Program Manager, Civilian), Col. Kevin B Peterson (Principal Deputy Program Manager, Military)
- Program Overview:
  - The Background and The "Family of Vehicles"
  - An illustration and diagram of the MRAP Enterprise and Structure
  - Team Descriptions: Acquisition, Business and Finance, Contracts, Engineering, GFE Integration, International Programs, JPO Forward, Key Partners, Logistics, Product Management, Production and Quality, Safety, Test and Evaluation
- Program Direction: Statement and full description of:
  - Mission
  - Purpose
  - Core Values
  - Vision
  - Forging Ahead
- Strategic Management: Targets and Action Plans using a Three-Phase approach (Phase I, Phase II, Phase III)
- Leadership: Leadership's Goal and Leadership's Objectives (Phase I, Phase II, Phase III)
- Systems and Processes: Goal and Objectives (Phase I, Phase II, Phase III)
- External Relationships: Goal and Objectives (Phase I, Phase II, Phase III)
- Back Cover:

> The MRAP has been phenomenal. What has been equally impressive is the breathtaking speed with which the MRAP has been fielded. I can't recall in my lifetime a major program that went from conception to fielding in light speed the way MRAP has, and it has been a tremendous asset...If given a choice, soldiers always choose to travel in an MRAP.
>
> **– Gen. David Petraeus**

- NOTE: Each printed Strategic Plan was accompanied by an 8 ½" × 11" Strategy Map illustrating, on one page, the linkage of all objectives and tactics through the goals of each focus area of the vision.

By 2011, the MRAP Program accomplished all of the goals, all of the objectives and all of the tactics published in the 2008–2010 Strategic Plan. The MRAP Program was now in transition. By 2011, there were 26,882 vehicles acquired, integrated and deployed, which far exceeded the original plan for 25,700 trucks. At that time, $44 Billion had been spent with another $11 Billion being planned for ongoing maintenance, sustainment and transition of the MRAPs to the ownership and control of each of the forces. The money to date was spent purchasing, preparing, deploying and maintaining the MRAP fleet. The average fully burdened cost for each vehicle came to $1.6 Million (i.e., $44 Billion/26,882 Vehicles = $1.6 Million per Vehicle) with an average Warfighter seating capacity of six, the military was spending $1.6 Million to protect $3 Million of potential lives-lost (Note: Although a bit morbid, the replacement for 1 Warfighter at that time cost the military $500,000. Six Warfighters amounted to $3 Million.). Financially, this is considered very Affordable, and in terms of human life alone, HIGHLY Affordable for Warfighter Lives, Warfighter families and the loved ones of the Warfighters!

With the MRAP Program in transition, Program Manager Paul Mann was succeeded by his Principal Deputy Dave Hansen, and a new Strategic Plan was necessary to guide and direct the program on this new leg of the MRAP journey. After the MRAP concept inception by Robert Gates in 2006, and the first four years of leadership by Paul Mann (2007–2010), Dave Hansen too the helm for the 2011–2014 timeframe. As the program transformed from a production, integration and fielding entity to a maintenance and sustainment function, it was eventually necessary to prepare for an evolution that transferred the vehicles and transition ownership to the respective military service led programs (i.e., Army, Navy, Marine Corps, Air Force and Special Operations Force). The contents of the full color MRAP 2011–2012 Strategic Plan included:

- Cover: "Joint Mine Resistant Ambush Protected (MRAP) Vehicles Program Strategic Plan 2011–2013" with Warfighter images, MRAP picture, MRAP logo and 'challenge coin' graphics.
- Forward: Written and signed by the new Program Manager Mr. Dave Hansen.
- Table of Contents:
  - Program Overview
    - Foreword
    - Background
    - Family of Vehicles (FoV)
    - Purpose, Mission and Vision
    - Guiding Principles
  - Forging Ahead
    - Key Success Factors
  - Strategic Focus
    - People
    - Process
    - Product
  - Acknowledgement
    - Key Contributors
- Program Overview
  - Background: Description of the beginning, the initial fielding and deployment efforts, the current state, and the near future state of the program.
  - Joint MRAP Vehicles Organization: Summary of the program focus, the product teams and the leadership team.
  - MRAP Family of Vehicles (FoV): Described and illustrated the four categories of vehicles (Category I: Urban combat operations, ambulance.

Category II: Multi-mission operations, convoy lead, troop transport, ambulance, utility vehicles. Category III: Mine/IED clearance operations, explosive ordinance disposal. M-ATV: Combat operations in rural, mountainous, and urban environments.), with photographs and groupings (i.e., Base Models, Enhanced Protection, Enhanced Mobility, Ambulance, Special Applications) with almost 30 specific configurations and variations.
- Purpose: "We protect those who protect us!"
- Mission: "We deliver MRAP vehicles, maintain the operational relevancy of the MRAP vehicle fleet, and sustain maximum readiness for U.S. and Coalition Forces."
- Vision: "We are a cohesive, people-oriented, rapid-response, jointly coordinated program, focused on new technologies, organized and coordinated to efficiently provide effective capabilities to Warfighters and customers."
- Guiding Principles: We value ...
  - Saving Lives
  - Stewardship
  - Transparency
  - Innovation
  - Collaboration
  - Perseverance
- Guiding Principles: We are...
  - **M**ission Focused
  - **R**esponsive
  - **A**daptive
  - **P**eople Oriented
- Guiding Principles: Secretary of Defense Robert M Gates, in a speech to the workers in the MRAP Vehicle Integration Facility, SPAWAR, Charleston, South Caroline, January 18, 2008

The war effort of the 1940s mobilized the entire American economy. That is not the case today, and while that may make yours seem like a lonely task, it only underscores how important that task is. Back then, President Roosevelt said: 'We must raise our sights all along the production line. Let no [one] say it cannot be done.' Those in the MRAP program have shown that it can be done. So keep raising your sights. Keep those vehicles rolling off the line. Your efforts are saving lives.

- Forging Ahead
  - Paul Mann, December 2010 (Joint Program Manager 2006–2010):

The Joint MRAP Vehicles Program has exceeded all expectations. We have delivered over 27,000 survivable MRAP vehicles faster than many thought possible. Thousands of lives have been saved, injuries reduced, and our Warfighters returned to serve another day. We are proud and humbled that MRAP vehicles safeguard our nation's greatest treasure. Mission success has positioned us for our next challenge, improving and sustaining these vehicles in the most cost effective manner. A cost control-conscious culture must take root and we must be effective advocates for affordability in every action we take. We all benefit from an affordable, survivable MRAP vehicle fleet.

- Excellence in Acquisition, Mobility, Logistics and Threat Protection.
- Keys to Future Success:
  - Workforce depth, alignment of goals, and career development: Goal #1—Personnel and Human Capital
  - Integrated, efficient, mature business processes: Goal #2—Improved Business Processes
  - Product focused roles and responsibilities: Goal #3—Enhanced Product Focus
  - Affordable, operationally relevant, Service supportable vehicles: Goal #4—Continual Product Improvement. Goal #5—Total Package Fielding and Sustainment

- Strategic Focus Areas
  - People: Goal 1—Recognize the contributions and perseverance of our skilled workforce and support their professional development.
    - Objective 1.1: Meet training and certification requirements of workforce employees' home organizations. Measure of Success—Percent of workforce meeting measured requirements.
    - Objective 1.2: Mentor and advise workforce employees on promotion opportunities and career advancement. Measure of Success—Percent of workforce mentored.
    - Objective 1.3: Provide developmental opportunities. Measure of Success—Percent of workforce taking developmental assignments.
    - Objective 1.4: Recognize individual, team, and organizational performance through awards. Measure of Success—Regular employee recognition ceremonies.
  - Process: Goal 2—Establish an integrated family of processes enabling the product and functional managers to plan, execute, monitor, and efficiently deliver emergent and enduring capabilities to the Warfighter.
    - Objective 2.1: Collaboratively review and inventory codified processes and standard operating procedures across the program. Identify critical gaps and improve linkages among processes and priorities to enhance integration of business systems for delivering capabilities. Measure of Success—Establishment of process linkages, closure of gaps resulting in more visible and predictable process performance.
    - Objective 2.2: Refine processes to gain efficiencies, and emphasize affordability and cost control as decision criteria. Measure of Success—Improvement of JMPV efficiency and affordability metrics.
    - Objective 2.3: Continually evaluate processes against Key Process Indicators and seek improvements. Measure of Success—Improvement of process Key Performance Indicators.
  - Product: Goal 3—Enhance unity of effort across JPO MRAP Enterprise by implementing Product Focus to our organizational concept, structure, manning and processes in order to more effectively support the Warfighter and the enduring Service requirements.

- Objective 3.1: Review and evaluate the current JPO organization, and implement a more unified, product-focused concept and structure supported by well-defined roles and responsibilities. Measure of Success—Achieve and document the approved organizational concept and structure, with defined roles and responsibilities.
- Objective 3.2: Identify, evaluate and track critical, product-focused, enterprise-level performance metrics (aligned with process metrics identified under Goal 2). Measure of Success—Institutionalize a JPO metrics process, that identifies, measures, and documents critical, product-focused, enterprise-level performance metrics, and is flexible, repeatable, maintainable and executable.

• Product Goal 4: Continually improve relevancy to the Warfighter by aligning vehicle configuration with enduring and urgent requirements, while standardizing capability.
  - Objective 4.1: Develop and implement comprehensive, cohesive and effective Fleet Management Strategy to enhance unity of effort and alignment with the Services' Tactical Wheeled Vehicle Strategies and enduring requirements. Measure of Success—Approval of Comprehensive/Cohesive/Cost Effective Fleet Management Strategy.
  - Objective 4.2: Maintain relevancy to the Warfighter by integrating new capabilities in vehicle configurations consistent with urgent and enduring requirements. Measure of Success—Percent of Vehicles relevant to Warfighter needs exceeds 95 percent.
• Product Goal 5: Execute a holistic, affordable, seamlessly integrated total fielding package and sustainment strategy.
  - Objective 5.1: Equip Warfighters via a comprehensive total package fielding strategy that ensures the force is trained and equipment is maintained and sustained to affordably enable mission accomplishment. Widely communicate milestones/critical path event and monitor the strategy in cooperation of the Services/SOCOM. Measures of Success—Timely completion of a Plan of Action and Milestones for this objective; military threshold readiness levels at 90 percent or higher.
  - Objective 5.2: Demonstrate affordable supportability by meeting criteria to successfully complete Department of Navy Integrated Logistics Assessment (ILA)/Army Sustainment Readiness Review, or the equivalent for other Services. Measure of Success—Successful completion of ILA/Readiness Review.
  - Objective 5.3: Meet the requirements for safety, suitability, reliability (performance), and logistical support via appropriate Joint reviews or Service equivalent (e.g., Full Material Release FMR). Measure of Success—Safety, suitability, reliability and logistical performance improvement; FMR or similar approval.
  - Objective 5.4: In collaboration with Service/SOCOM, develop appropriate plans to affordably support recap/reset; retrofit; redistribution; and long-term storage. Measure of Success—Plans in place for each component.

- Objective 5.5: Enhance sustainment level battle damage repair capability forward. Measure of Success—Improved Battle Damage Assessment and Repair throughput; improved Mean Time to Repair.
- Success Quotes on the back page and within the plan.
  - Back Page:

> There isn't a trip I've taken into [Iraq and Afghanistan] where some young soldier or Marine won't say to me, 'keep those MRAPs coming. They save our lives.' [MRAPs] saved countless lives, and believe me, there's not anyone on the ground over there who understands the threat that doesn't know that's exactly the case.
>
> **—Admiral Mike Mullen, Chairman Joint Chiefs of Staff (June 3, 2010)**

  - Back Page:

> I know for a fact if it had not been for these vehicles I would have lost some Marines and had worse injuries to them and to myself. We had 38 Marines hit by IEDs. Out of those, some were even hit twice, and 10 Marines were hit three times. Again, with no loss of life. It is amazing how the MRAPs kept us alive and allowed us to do our dangerous job just a bit safer. The best way to explain how great these MRAP vehicles are is this: My Marines and myself received medals [purple hearts] for our injuries, instead of our families receiving American flags.
>
> **—Gunnery Sergeant David Pizana, 8th Engineer Support Battalion, Security Platoon (4 May 2010)**

  - Within:

> I was involved in an IED attack March 27, 2010 in Zhari District, Afghanistan in an RG – 31 model. I wanted to express my thanks, and appreciation in providing these great vehicles. If it wasn't for this vehicle, all of us in this MRAP would have been killed, or severely injured. You guys should be proud, because you too are serving our great country, and doing such a fantastic job, increasing our chances of surviving IEDs.
>
> **—SPC Michael Fleury, A. Co 1-112 Infantry Regiment, 4th Brigade Combat Team, 4th Infantry Division (21 May 2010)**

As 2013 closed, and 2014 began, the MRAP program was moving about half the vehicles to long term storage and transferring the ownership of the active vehicles to the major Services and other Customers (domestic and international). The Strategic Plans served to communicate to all of the U.S. Military Services, the contractors supporting the war efforts, The President, Congress and the Public the MRAP 'Program Message' and the program's direction, alignment motivation, execution and results. The plans themselves did not create success, the people involved in the acquisition, integration, deployment, use, utilization, maintenance and sustainment did. However, the plans did serve as "The Message" for direction, alignment, motivation and execution. After the withdrawal from Afghanistan in 2021, and 20 years of war, combining the fatalities of both fronts, 6,892 lives were lost. Although it's quite difficult to estimate the number of deaths avoided, the statistics show that the number of IED deaths dramatically declined after 2010 when the MRAPs were completely fielded and in full use (NOTE: From 2003–2010, the number of IED deaths when charted, formed an accelerated parabolic curve, and when extrapolated, the curve showed the rate of death increasing with a prediction of several thousand more deaths that didn't happen.).

A STRATEGY TO REVIVE A GIANT (STRATEGY BEFORE TACTICS)

One of my first projects during the beginning of my consulting career in 1995, involved the creation of a strategic plan to get AT&T back into the local consumer and commercial communication business. I was invited-in to participate with a team of strategic marketing people in AT&T Atlanta office for the purpose of drafting a ten-year plan focused on returning AT&T to the position it once held prior to the Breakup of the Bell System in the 1980s, establishing a new prominent role in the telecommunications industry. The customer targets included all types of individuals and organizations requiring communication means and devices for all types of voice, data and images, both digital and non-digital.

At the beginning of the next year, the deregulation of the industry took place by way of the Telecommunications Act of 1996. This put the wheels in motion for the effort. The targeted partnerships, acquisitions and mergers started. Initially, the moves were small, specific and targeted. Then in 1997, when IBM Executive C Michael Armstrong was hired with the vision of a global telecommunications supermarket, a clear direction emerged and emphasis was placed on internet services in the dot-com industry as well as all types of digital media (e.g., TCI, MediaOne) and it included general customer communication capabilities.

In 1998 more takeovers, alliances and purchases took place (e.g., Teleport). And, in 1999 IBM's Global Network business was purchased along with the Olivetti & Oracle Research Lab including many other strategic transactions. For the next several years, the investments and industry footprint growth continued. In the final year of the plan, although it was not written in the original plan, a reverse merger of AT&T and Southwest Bell occurred which relocated the AT&T Corporate Headquarters in Dallas Texas. Today, 20 years later, and 150 years after Alexander Graham Bell founded the Bell System, AT&T Inc. operates as a global telecommunications company.

CRITICAL LINKAGES

For most cases, the linkage between Strategy and Tactics is determined and defined from Strategic Goals to Operational Objectives to Tactical Targets using both quantitative and qualitative metrics to measure performance. It's a connection of, "Where are you going? What are you doing? How are you getting it done? Who is serving the role and responsibility for results and accomplishments?" (Strategic ⟷ Operational ⟷ Tactical ⟷ Personal). Strategy directs Operations. Operations align Tactics. Tactics motivate People. The Enterprise area is Strategy, the Systems zone is Operations, the Process(es) region is Tactics, and the Procedures vicinity is People. If you use an Affordability Scorecard (Customer, Business, Process, People), Customer Satisfaction Enterprise Strategy links to Customer Operations Systems (to Meet Requirements, Ontime, Affordable Cost) links to Tactical Processes (Takt Time, Quality Standards, Efficient Cost), links to People Procedures (the Work, the methods, the means, the tools, etc.). As performance is measured, each metric and measure eventually links to the Goal(s). The comprehensive view of the metrics and measures of the baseline, goal, objective, current actual measurement illustrates the pathway of performance from the beginning to the present, to the planned future.

YOUR PATHWAY AND YOUR JOURNEY

Whether it's a personal strategy, a team strategy or organizational strategy, once established, it provides a path to take from where you are (your current state) to where you're going (your future state). Lewis Carroll once said, "If you don't know where you're going, any path will get you there." Once you've determined where you're going (The Goal), the path must be defined to get you there (The Strategy). The strategy and strategic plan guides, aligns and motivates your actions and activities, as well as the actions and activities of the people.

Don't be surprised, if during the journey, it becomes necessary to adjust the plan and pathway. The strategy may stay the same, but the plan may change. And be aware, there are those unique situations where the strategy may have to change due to unforeseen circumstances that arise, causing a shift in the route or an alternative course. For example, I know an individual who had a goal of studying animal science to become a veterinarian. While in college, she was exposed to veterinarian medicine through various courses at the university. A great deal of veterinary practice involves the euthanasia of pets. This unforeseen and unexpected exposure caused her to shift her path. She eventually received her B.S. in Animal Science and eventually went on to get her Master's Degree in the same discipline. Although not a veterinarian, her work in animal science has provided a pathway to work with veterinary medicine and medical devices. Most recently, her strategic plan had to change as a result of having her third and fourth child. Currently, with a career of "motherhood" (now her second career), her pathway, for the near term is apparent.

For your pathway, I suggest applying the ten points of leadership discussed in the previous chapter. Specifically, Action #5 "Executing the Plan" is one of the leadership critical actions that is sometimes overlooked by managers, the executive and team leaders. For this to take place, a strategy and plan must be present. Define your strategy and create your plan. Begin your pathway taking the first step of your plan. Continue, adjusting when necessary. You may find that over time, the plan can be extended, changed, renewed or rewritten.

5 Enterprises and Systems

Some people regard private enterprise as a predatory tiger to be shot. Others look on it as a cow they can milk. Not enough people see it as a healthy horse, pulling a sturdy wagon.

– Winston Churchill

If you do something every day, it's a system. If you're waiting to achieve it some day in the future, it's a goal.

– Scott Adams

ENTERPRISES AND SYSTEMS OF AFFORDABILITY

An enterprise is comprised of systems that deliver value to customers at a price the customer can and is willing to pay at some profitable price/expense level for the organization. Often, businesses use the traditional approach and framework of thinking for managing the business using the factors of Supply, Demand and Price. Many managers have been taught to manage their business by attempting to manipulate each variable in a variety of ways. For example, to address supply one makes more product or less product, and for services provide either more or less. For demand, either advertise the product or service through marketing, and the often use an attempt to increase purchases through sales programs or sales incentives. In terms of price, either raise it or lower it. From an enterprise of systems perspective, supply, demand and price are lagging indicators of performance. Supply, demand and price are the results and outcomes of the systems in operation. If there's a problem supply, demand and price are symptoms to address, not the root cause of the problem. Affordability has shown that one can increase demand by increasing the speed and availability, improving quality and decreasing the cost by reducing the expense with the ability to the lower price; hence, the product or service becomes more attractive for the customer. As for supply, increase the speed of production and availability with the added benefit of flexibility to decrease the flow when necessary. Price is a factor of the expense of providing the value and the customer's willingness to pay (unless the price is controlled and regulated as it is in China or subsidized like it is for the Affordable Care Act in the US). In fact, many say that an enterprise of socialism does not work but fail to realize that it does work as long as there's an outside source subsidizing the systems (Note: Ultimately, all socialistic have failed.). As an example, I learned, growing up in south Florida, that Castro's effort in Cuba was initially successful because it was subsidized by billions of Russian Rubles. Eventually, Cuba failed when Russian socialism began to fail and their outside subsidies stopped. The American affordable capitalist economy succeeds because it operates on a sustainable engine where demand grows when organizations provide value at a price that generates revenue greater than expense and cost.

DOI: 10.1201/9781003772934-5

In most cases, an enterprise, by definition, is a business or company. An Affordable enterprise is described in a much broader scope, and it entails the delivery of value to a customer using systems comprised of processes (and support processes) that provide product(s) and/or service(s) to a person or organization that is willing to pay the designated price. Those systems contain processes that are the methods and means by which the customer receives the product(s) and/or service(s) provided. Too often, organizations are myopic in such a manner that their focus is within themselves and fail to fully understand and appreciate their suppliers, and even at times underappreciate their customers. Management sometimes misplaces emphasis on the internal rather than both the internal and external. Affordability is about the whole enterprise "from raw material to finished goods delivery" or "from the entire pre-prep to the delivery to the completion of the transaction." In other words, everything upstream, in the stream, and to the end of the downstream is the enterprise.

I learned from the Toyota Production System, that three of the key components of every successful enterprise is the understanding of Value, Value Stream and a Value Stream Map (i.e., The Value Stream Map is a diagram that illustrates how an enterprise flows and fulfills demand, from very beginning before the creation of value, to the very end, after the creation of value to when the value is delivered to the customer.). When studying a process, the basic structure is often referred to as a S.I.P.O.C. **S**uppliers → **I**nput → **P**rocess → **O**utput → **C**ustomer. When processes are linked, combined and integrated, they form a system. When all systems are conjoined to create and deliver value, they are an enterprise under the paradigm of Affordability.

AFFORDABILITY BARRIERS

There exists a multitude of hurdles, roadblocks and barriers for anyone traveling on the journey of Affordability. Organization history, marketplace traditions and competitor challenges are all examples of external obstacles. Most often, the worst obstructions come from within. I once heard Dr. Deming say,

> Leaders too often blame the people for the problems of the organization. Most often, the people of the organization aren't 'broke', the system is broken. Fix the system. Use the people to fix the system and solve system problems using the people's solutions.

Additionally, organization culture does offer severe hinderances, impediments and difficulties. The organization's cultural issues are once again not the people's fault, but in this case, it's leadership's fault. Leadership must transform the culture by instituting multiple changes directed at achieving the desired future state, while aligning and motivating the people to adopt and adapt to new values, principles, beliefs and systems. Management is about 'things', leadership is about people.

Finally, to move from a traditional organization construct or an archaic outdated establishment to an Affordability 2.0 design, leadership must employ a vision, a message, direction, strategy, a plan, tools and resources for the people to embrace and inculcate into the culture.

LEADERSHIP AND CULTURE

As discussed in an earlier chapter, Affordability 2.0 Leadership is about actions (i.e., setting direction, aligning resources, motivating people, communicating 'the message', executing the plan) and behaviors (i.e., modeling the way, inspiring a shared vision, challenging the process, enabling others to act, encouraging the heart). The actions and behaviors of the organization's leadership will establish the culture, define the characteristics of the institution and its role in the enterprise, and determine the profile of how business is done and how work is accomplished.

Although leaders have influence on a culture, the culture of an organization will emerge and morph as time goes on, business practices are executed and the people within the enterprise operate and interact. Events and activities shape the "personality" of the 'outfit', and the culture evolves through planned and unplanned proceedings and results. In other words, just like a human growing to maturity, there are stages in life that shape and manipulate what that entity will be.

Leaders can intervene when things are going wrong or need to be adjusted. Leaders can institute change and transformation for new direction. Leaders can maintain good 'habits' and quelch bad ones. Leaders can continue in good direction, start new direction, and stop going in a bad direction. It's leadership's responsibility to see to it that the establishment's culture is aligned with their place in the enterprise using excellent systems to create, produce and deliver Value in accordance to Customer Requirements at a price the Customer can, and is willing to pay, at the same time, managing expenses such that a profit is made.

GOVERNMENT IMPACT

I have three words to describe why unaffordable conditions, created by government, negatively impact, and destroy Affordability: Waste, Fraud, Abuse. These three words are valid for private institutions as well as government. Remove waste fraud and abuse, and systems will operate more efficiently and effectively. It's a 'no-brainer'. However, since WWII, and for all of my life, I've seen waste, fraud and abuse of US wealth get directed and squandered on political intent, "good will," and bias focused that mostly ended up as "fat bacon and pork ladened programs" that yielded little for the people but achieved much for the politicians involved. It is true, efforts in civil rights, improvement of highway and airway travel, advancements of technologies and numerous discoveries in the science of healthcare were a result of government involvement, yet they pale in comparison with the capital misused on all the other programs and projects using taxpayer money. Today, affordability is a keyword thrown around to either attack or support political efforts. Getting to the bottom line, a great deal of waste, fraud and abuse is openly acknowledged. The key question: "What is the remedy?"

Overspending and inflation continues to run rampant today. The deficit continues to increase. Unless it is corrected, it is easy to predict a total melt down and failure of the United States of America. Looking through a lens of Affordability, the strategic solution is clear; increase demand of the products and services of American providers, while removing roadblocks and barriers that impede advancement, and execute a concerted effort to eliminate waste, fraud, abuse and the misuse of America's wealth. In other words, increase flow, decrease waste and eliminate unneeded programs. Increasing taxes on the citizens of this current system does not "increase revenue" as many ill-educated and ignorant politicians claim. And attempts at "increased taxing of corporations" (Note: Corporations don't pay taxes, People do.), or "taxing the ultrarich" (Who, when combined, can't statistically overcome the deficit and financial deficit shortfalls), only obfuscates reality and treats symptoms without solving the problem at the root cause (i.e., more money spent than "revenue" gained ... Non-Affordability for sure). Let's consider a different framework and format of thinking ... Affordability.

So, using the principles of Affordability, where can we focus? Here are a few areas of 'low hanging fruit' or quick wins that can be pursued.

- Regulations: Remove regulations that hinder and impact business success.
- Taxes: Lower taxes and use taxes judiciously. More money will not solve the problems at hand. "Throwing Money at the Problem" does not fix the problem.
- Debt: Reduce the debt by decreasing spending and eliminating waste, fraud and abuse. At the same time, use techniques to support interest in American products and services.
- Price Control: Don't! Let the customer and the market determine the price.
- Subsidizing Programs: Don't! If a project or program can't stand on its own, don't support failure.
- Political vs People Initiatives: Although redundant, our government should be executed by the people and for the people with a focus on freedom, rather than operate as a government defined by politicians and for politicians with power and control of their position as the central theme.
- Immigration: Yes, immigration is a good thing. Recent mass immigration without boundaries have failed and produced more waste fraud and abuse (Results: higher crime, misuse of funds, waste fraud and abuse of taxpayer money). In the past, merit-based immigration and the factual need and demand for human labor and intellectual involvement was a policy that succeeded (Note: All of my great grandparents were immigrants of the last half of the 19th century).
- Environmental: Clean Air and Water is a desire for everybody. Not government environmental schemes to financially line political pockets
- Cultural and Social: The American experiment is clear, assimilate and embrace the existing principles set in place by the majority, not the minority. Respect cultural and social minorities (Note: "Minority" is NOT used here with racial connotations) but, do not legislate and divide the people using every principle from every facet of the population (Remember: "... of the people and by the people ..." does not mean "... of every person and by every person ... " We are a Democratic Republic.)
- Crime and Safety: Preserve, Protect and Defend the People. After all, we are a lawful country.

- Picking and Choosing Winners: Let the marketplace choose. When there was a push by government during the late 20th century to "off shore" products for price and price alone, we eventually suffered as a country and ultimately suffered when jobs went away, availability was impacted (e.g., COVID) and the price eventually increased.

In addition to the focal points mentioned above, we must recognize another "elephant in the room." Today, propaganda runs rampantly seeking to destroy political competition. Facts are fabricated, feelings are seen as foundational fundamentals, information is invented and science is slanted in an attempt to build a case for a cause without regard to honesty, integrity and truth (Note: The values of Affordability include all three). It has become difficult because lies and liars are not only abundant, but false accusations of others often go unchecked. Although Affordability is based on actuality, reality and true fact, society must contend with these ills, and vanquish falsehoods.

Somewhat related to propaganda, transparent communication, a tenet of Affordability, is not an honored attribute of most politicians. Half-truths, lies and secrets supporting their causes and power seem to be their modus operandi when speaking to the people. It's understood, within any society, politics, dogma and doctrine will exist, but, when yielded for personal power and gain, at the detriment of the people, Affordability will fail.

And, in terms of failure, why are so many big cities unaffordable (i.e., New York, Chicago, San Francisco, Los Angeles, Philadelphia, Portland, Seattle, Boston), while other cities around the country are affordable and growing (i.e., Houston, Phoenix, San Antonio, Dallas, Miami, Jacksonville)? The answer is just plain common sense: Affordable cities tend to be more people centered, while unaffordable cities are more politician centered where power of position and control of the people reins.

With all that said, there is a 'flip side' where government actually and positively can influence strength, growth and success in an Affordable manner. The Victory of WWII and the following boom of the 1950's, the "Space Race" of the 1960's, Equal Rights of the 1960's and 1970's, the 'economic and political turn around' in the 1980's, the balanced budget of the 1990's, the response to terrorism in the early 2000's, etc., are all examples positive impacts of government.

Affordability flourishes when a government by the people and for the people based on freedom is present. Not a government that is defined by politicians and for politicians to control the economy and the people (that system ultimately fails as does the government). The best government for Affordability operates in a manner as to preserve, protect and defend its people in terms of freedom. Not a government that preserves, protects and defends the power and control politicians have over the people. An Affordable Government is one that seeks to eliminate waste, fraud and abuse, and is one that removes barriers roadblocks and hurdles that stifle individual and organizational success, and is one that operates in a manner that abstains from picking winners and losers using subsidies, and is one that preserves, protects and defends the people by providing safety, security and safeguards success from the individual to the enterprise.

SELECT PRODUCT AND SERVICE ENTERPRISES

Within the US, there are a multitude of specific enterprise areas that would benefit from Affordability. Most of these enterprises are commonly used by today's media as examples of unaffordability. All of these enterprises serve individuals and households in the US in one way or another. In 50 years, 1976 to 2026, individual income rose from $14,960 to $89,500 and household income increased from $12,686 to $70,000 (Note: 600% and 550%, respectively). From 1976 to 2026, the US GNP grew from $1,656,000,000 to $24,058,900.00 (Note: 1,453%). Save it to say, we've squandered a lot of wealth in the US during the past 50 years. The enterprise categories below represent a great deal of the economic drivers that are of concern to everyday individuals. If Affordability gained momentum in these areas, great wealth and prosperity can be gained on both the individual or consumer buying side of the coin, with profitability and prosperity on other side being. Enterprises would be wealthy and the delivery of Value would be much greater. Here are 11 examples of economic areas of Affordability:

Healthcare: In 1976, the average doctor's visit cost was $15–$25, while the average annual healthcare cost amounted to $638 per person. Fast forward to 2026, where the average physician visit is at least $80 and healthcare is above $15,474 (Note: last best estimate 2024) per capita. This is an increase of 400% and 2,425% respectively. Applying Affordability of faster (i.e., more access) and better (higher quality) healthcare at a lower cost would go a long way to achieve Affordable Healthcare (Note: You may refer to my last publication – "Healthcare Affordability"). In a pilot program in the state of Georgia, I was partnered with the IISE (Institute of Industrial and Systems Engineers) on a project sponsored by the GHA (Georgia Hospitals Association) targeted at improving the speed and quality at a lower cost. Over a several year period, we certified 74 Lean Six Sigma Black Belts across the state representing a great deal of the state's healthcare institutions. Each Black Belt Candidate was required to execute a project after training and knowledge transfer to demonstrate competency and capability. In cost reduction results alone, $57,409,457.00 was saved for an average savings per project of $775,803.47 (Note: each project savings amount ranged from $8,700.00 to $8,754,00.00). In addition to cost reduction, each project was evaluated on increased process speed and performance, as well as better accessibility with improved quality and excellent system performance.

Housing: As it was in healthcare, the housing industry and market increased in cost and decreased in accessibility. In 1976, those of an average household income of $12,686.00 were purchasing $48,050.00 homes with a 7.7% mortgage (Note: The household income was ~25% of the housing price). Coincidentally, I purchased my first home in 1976 for $29,000.00 on a salary of $9,800.00 (Note: I stretched it at ~33% of my income). Fifty years later, those with a mean household income of $70,000.00 are purchasing $420,000.00 homes with a 6.3% mortgage (Note: The household income is

~17% of the housing price). If anything, from this data, it appears that the "Affordability" of homes is about the same, or better, in 2026 as 1976, with a slightly lower mortgage rate (Note: Consideration of all household expenses much be considered.). Apparently, in a general sense, home affordability is as it was in 1976, however the data is quite skewed when it comes to urban, dense city locations. In such a case, Affordability should be applied to the areas in most need. Affordability can also be applied to every home provider to increase speed, improve quality and lower cost to increase demand, foster growth and enhance capacity.

Insurance: Affordability has been applied in commercial insurance to decrease risk, lower premiums and improve operations of policy holders. Working with CAN out of Chicago Illinois, Lean Manufacturing methods were used with a client who manufactured tractors. When Lean was applied, injury claims decreased, speed of manufacturing increased and manufacturing cost went down. Insurance companies can partner with clients to apply Affordability techniques to increase process speed, improve quality output and lower cost, thereby increase profits and demand.

Higher Education: Over the last 20 years, I've been delivering a program to certify college students using an accredited program from the IISE focusing on the Lean Green Belt BOK (Body of Knowledge). Serving more than 20 Universities, and 1000's of students (Undergraduate and Graduate) from a variety of disciplines, graduates were able to take the Lean Toolbox into their workplaces and apply the methods for improvement of systems and processes. Since I was one of several professionals participating in this program, literally 10's of 1000's of engineers, business majors and other disciplines have taken this knowledge and capability into numerous economic sectors.

Food and Nutrition: This is a hot contemporary topic that has emerged due to negligence in the industry for the past 50 years or more. Tightly linked to healthcare, food and nutrition is a topic that greatly impacts "kitchen table" issues. It's estimated that an average family of four spends $12,000.00 to $20,000.00 annually on food (Note: It was estimated at around $1,560.00 in 1976). The nutritional content of our food is estimated to be much less now than it was 50 years ago, with more preservatives, sugars, salt and chemical additives. Fast Food is cheap but low in nutrition, as are chips, crackers, candy, cookies, donuts, sodas, sports drinks, sugar-added fruit juice, store bought cakes, 'instant' meals and numerous other synthetic and artificially created "food." Affordability applied here would focus on content, quality, cost and accessibility.

Energy: Included with technology, energy drives the economy as well as wealth, prosperity and Affordability of US goods and services. Using Affordability to remove wasteful policies that have plagued the industry for years and increasing the availability of power to enterprises and individuals alike, the mantra of faster, better, more affordable means growth and prosperity. Within Affordability, safety is a key component of any and all programs in this realm. The safest and cleanest forms of energy are nuclear and renewables (wind and solar), and if Affordability was applied to increase efficiency and effectiveness, a better multi-sourced solutions can be offered at lower cost. The often stated the all-of-the-above approach can be leveraged if Affordability was applied to all sources of energy.

Manufacturing: The enterprise most associated with Affordability is manufacturing. However, since Affordability is not utilized universally, there still exists a multitude of locations that could leverage this 'science'. Often, when I'm invited in by a potential client, the presence of Affordability is nowhere to be seen. Whether it's a light assembly production line of small handheld devices, or a heavy industrial construction of a building, there typically exists a lot of waste in terms of time, material and resources. A great deal of my application of Affordability involved products from small steel hooks for use with lifting equipment to large fighter jets such as the F/A-18 Super Hornet. Manufacturing is probably the most advanced enterprise in terms of using Affordability.

Transportation and Distribution: With the reality of our worldwide economy, the enterprises of transportation and distribution are fully entrenched globally. Faster, better, more affordable resonates within these enterprises. The speed factor, the quality dimension and the efficiency component for moving items domestically and internationally is a formula for performance excellence in all supply chains.

Retail: With margins being very slim in this industry, Affordability is best applied to primarily reduce expense and spend and secondarily improve service quality, customer demand and fulfillment requirements. Whether the retail segment is internet purchasing, department stores, specialty stores, mass merchandizers, big box stores, supermarkets, convenience stores, restaurants or even hotels and motels, Affordability strategy, systems and structure can be readily applied to large, medium and small establishments

Military: With several military applications over the past 20 years, the Affordability philosophy and model is a good fit for all service branches. From the F/A-18 Super Hornet Program to the MRAP Program, recent history has demonstrated the military is a sound field of operation for Affordability methods and strengths. The Super Hornet program reduced the per plane cost by $5.1M, increased the production capacity from 42 to 67, and increased the quality from around 4σ-5σ to 6σ+, and grew the market demand for the aircraft from the Navy and USMC only to also include 4 other countries (i.e., Australia, Japan, Canada, Switzerland). The MRAP deployed more than 25,700 trucks in three years (i.e., the fastest fielding of a ground combat system in the history of the world).

Government: To finalize my compare-and-contrast approach, from 50 years ago to today, the GNP went from $349,400,000,000.00 to $7,010,000,000,000.00 increasing 2,000%. Spending was $430,145,000,000.00 in 1976 compared to $7,670,000,000,000.00 last year with an increase of 1,783%. 50 years ago, revenue was $299,400,000,000.00 to a last year amount of $5,230,000,000,000.00 with an increase of 1,750%. The standing debt is now more than $106 TRILLION and growing. What seems to be obvious over this span of time is the constancy that spending constantly paces ahead of revenue and the GNP. The country's overall potential wealth and prosperity continue to grow, while the waste and deficit spending in government continues to occur, and the citizens continue to bear the burden of debt. Applying Affordability, the elimination of waste, fraud and abuse of the system, continued reduction of operating costs by increasing system efficiency and effectiveness, increase the accuracy of the services applied by improving processes and eliminate programs that are ineffective. "Easier said than done" is an appropriate response. If we continue this path, the journey will end in failure. If we apply the Affordability Formula, better days are yet ahead.

Universal Economic Formula: Although not every enterprise category in existence is represented here, there is a great enough sample for proof of applicability. I've only included enterprise areas that I've worked within directly and have Value Stream experience. Affordability has proven to be a Universal Economic Formula serving as a foundation and framework for increasing performance and leading transformation. It takes enterprise thinking to apply this formula and create successful systems. Please keep in mind, Affordability for the individual is not a single enterprise perspective but an overall economic view of the purchasing power of every person. Affordability Thinking must encompass all Enterprises, and should be a theme for every industry to bring more profitability and prosperity to US Citizens and US Corporations, not to mention US Government.

ENTERPRISE THINKING SUCCESS STORIES

Enterprise thinking is about the consideration of big picture philosophy. From suppliers of suppliers to customers of customers. Toyota is a good example of this type of thinking. Toyota's most often delivers Toyota vehicles to Toyota dealerships who in turn sell those vehicles to customers (i.e., the 'end user'). Toyota suppliers receive component parts from their suppliers (e.g., a muffler supplier receives the muffler metal from a metal supplier), and deliver material, components and parts to Toyota manufacturing plants. Connecting these entities illustrates a simple "big picture" of the Toyota Enterprise:

Sub-Supplier→Supplier→Toyota→Dealerships→Customer

This big picture view serves as a grand Value Stream that can be expanded into a more detailed map of the Toyota Value Stream and the Value Streams of the Sub-Suppliers and Suppliers, as well as the Toyota Dealerships. Such a perspective conveys where everything fits, who is involved at every point along the Value Stream and can be used to leverage Affordable practices (e.g., The Toyota Camry Plant in Kentucky and the Toyota Tundra and Sequoia Plant in Texas may use such intelligence to leverage component parts purchasing from suppliers).

MANY LOCATIONS, MANY DIFFERENT PHILOSOPHIES

Often, as companies grow and expand, they acquire other organizations or merge with different competitors to accomplish their growth goals. An organization I've served, Crosby Tools, was faced with such a reality at the time I was invited in to provide Lean expertise and system improvement services. Crosby Tools, headquartered in Tulsa Oklahoma, had satellite production facilities in Longview Texas, North Little Rock Arkansas and Brampton Canada. Their primary business is manufacturing tools for the heavy lift industry (i.e., construction, oil production, mining, etc.). The plants worked with steel to forge parts for lift equipment (e.g., hooks for cranes), mill steel fittings and shape metal for other applications (e.g., clips, latches, blocks).

My first engagement was in their Longview Texas plant that forged steel using large hammers and dies to produce crane hooks for lifting heavy loads. The leadership of that plant had their "own" defined Crosby Enterprise and sought to create a Value Stream for use at the plant addressing their own products. Their enterprise involved raw material (steel) hammered and forged to produce a crane hook that had to be trimmed, finished, painted, packed and shipped per customer order. Accompanying the Value Stream Map, the leadership team created a spaghetti diagram to physically illustrate how material flowed in, the material was shaped into a product, painted and finished per specification, where the finished good was packed and shipped out to the customer. The enterprise view of the Value Stream and the physical depiction of the material/product flow illustrated by the spaghetti diagram was used to identify system defects for problem solving teams. Receiving rhythm of incoming steel, maintenance of steel hammer sleeves and dies, contamination of products, pacing of painted parts, packing and shipping issues were all addressed using this big picture approach to pinpoint and address complex problems with solution teams.

The next stop was the North Little Rock plant in Arkansas. The leadership team there was a bit reluctant to take the same approach as Longview. I often heard, "We're different here. That's not the way we do things around here." My role was to convince the top leader about the benefits we captured from the Longview activities and use those learning's plus all the unique learning's of the plant. Once the general manager was on board, the direction, alignment, motivation, message and plan fell into place. Using the Affordability artifacts created, we were able to employ problem solving team to attack and resolve issues within the system.

The third location was Brampton Canada. Not only a different plant, but a different country. The general manager was already onboard and ready, but a few of the leadership team members were resistant to embrace and adopt the two successful approaches we already had in Longview and North Little Rock. After drafting their VSM, we took a three-plant view of all three at once and noticed the commonality of all sites, with only small incidental differences. Some of their metal work involved forges, with about half of their business involved CNC carving and shaping parts. Accommodating those two streams, two maps were created that illustrated parallel core methods, with common flows from suppliers thru process to customers.

Finally, the flagship plant in Tulsa Oklahoma was the culminating site. Their products were like some of the other plants, plus they had the capability to forge much larger products than any of the other plants. Being a headquarter site, they had a bit of a superiority attitude over the corporation. The challenge here was to convince the leadership team that their Value Stream Map and Enterprise view was complimentarily similar to the other three plants. Once rendered, the leadership team used all four maps to integrate and amalgamate all three maps, taking into consideration the nuances that did occur at the other plants. This one view of the Enterprise and Value Stream was used thereafter the big picture that permitted everyone across the organization to understand their role and responsibility in creating and delivering value.

WHEN THE FINGER POINTING STARTS, AN ENTERPRISE PERSPECTIVE IS BEST

A little more than 20 years ago, when AT&T was expanding their cell phone communication footprint, and Lucent Technologies was building the nationwide infrastructure of cell phone towers for this effort, I was invited in to serve Anixter and their part of the endeavor. Anixter was serving this effort as the warehousing and distribution for electronic components, materials and parts for the construction of the cell phone towers. The business for Anixter amounted to hundreds of millions of dollars and created a corporate division for this program. Other entities involved included: logistics support, freight carriers, cell tower component companies, and cell tower construction contractors. To say the least, each factor and facet involved in the project was very self-centered and solely focused on their portion of the business. Every time a problem or issue appeared, the "finger pointing" started.

One particular problem that occurred ended up being quite costly to the parties directly involved. Each cell tower was equipped with an electronic cabinet where all circuit boards, batteries and electronics for every cell tower was located. All too often, the cabinets arrived damaged and eventually deemed unusable. Under the Lucent contract, Anixter and the cabinet supplier had to share the cost of replacement. The base cost of the cabinet to Lucent was $34,000.00 which meant Anixter was charged $17,000.00 and the cabinet provider was also charged $17,000.00. As expected, the finger pointing began. Anixter knew that from receiving to shipping, the handling of the cabinets used high quality processes that didn't damage freight. The cabinet supplier had a long-standing reputation for producing quality products. The two companies were at war over the issue and it seemed that the solution was out of sight. Other than the cost, the time impacted to replace the damage cabinet delayed the cell tower implementation, and the perception of customer quality for both parties was damaged.

This part of the Lucent Technologies program Value Stream seemed straightforward:

Cabinet Supplier → Anixter → Cell Tower Construction Site

However, as it's often quoted, "the devil is in the details." In order to get a detailed visual of this portion of the Value Stream, we filmed a cabinet being received, put away, inventoried, picked, packed and shipped at the Anixter Warehouse in Alsip Illinois. We visited the cabinet manufacturer in NC, who walked us through their manufacturing and shipping process. Subsequently, both teams viewed the Anixter video and discussed the process links from manufacturer to warehouse to cell tower construction destination. Both parties were convinced by this enterprise view that the issue was much deeper than Anixter's system or the cabinet supplier's production system.

Both parties collaborated to dig into the rest of the enterprise by researching the systems involved. To make a long story short, every process of every system being used was researched by the cooperative study team employed to solve the problem. The discovery amounted to the following: Lucent contracted a Logistics Company (that shall remain unmentioned), and that company contracted a cheap carrier in an effort to cut cost and increase margin. It was discovered that the cheap, lower quality carrier was sometimes damaging the cabinets on the way to an Anixter warehouse, and at other times, damaging a cabinet on the way to a cell tower construction site. Beginning with the enterprise view and digging into the systems involved, what seemed to be an unsolvable complex problem, was quickly resolved when the parties involved cooperated and collaborated on researching and resolving the matter.

The examples provided are but only two representations of the billions of opportunities that exist. Improving people, procedures, processes, systems and enterprises is the pursuit of continuous improvement that should be a constancy of purpose for all organizations. Affordability is about creating, implementing, maintaining and sustaining enterprises through continually improving the system and their related components. Every person's work is involved in at least one enterprise. Knowing the enterprise(s) you're involved with is critical to understanding what's required for your success.

YOUR PATHWAY AND YOUR JOURNEY

Think about the enterprise(s) in which you exist, the value you deliver and the customers you serve. For me, I'm a consulting enterprise, in the service business of delivering improvement solutions (Areas Include: Leadership, Strategy, Performance Excellence) to customers of all types who can and are willing to pay for change and transformation of their systems. Your Value may be product(s) or service(s) or both. You may be a primary provider or a component supplier or a value stream support entity. No matter whether you serve several enterprises or just one, intimate knowledge of the enterprise(s) you're in will assist you in navigating your journey. Satisfying universal requirements and specific customer requirements will help in establishing your own UVP (Unique Value Proposition).

Define your part of the enterprise from upstream suppliers through your value stream to downstream customers and ultimate end users. Detail the requirements you have for your suppliers and partners and define as detailed as possible customer requirements for your products and/or services. The next chapter provides some tools and techniques for designing Affordability for your systems, processes, products and services.

Determine the resources you will require to create, produce and deliver your Value. Understand the means methods and tools to efficiently and effectively deliver Value to the targeted Customers. Render the design and solution options that fit the Customer targets and target markets you've chosen. Pick an option to assess and evaluate. Run simulations, trials and experiments to test and evaluate your options.

When you arrive on the "right solution" (i.e., it's proven to be a "good fit" and ready for implementation), prepare and package the implementation for deployment. Once deployed, maintain and sustain the solution, but don't forget to pursue continual improvement and sustainment.

6 Design for Affordability

By failing to prepare, you are preparing to fail.

– Benjamin Franklin

DESIGNING FOR AFFORDABILITY

Whether you've been in business since the 1800's or have just recently started a business in the past year, many businesses fail due to lack of preparation or continued adaptability and flexibility of the business. Most businesses succeed due to a combination of strategic planning, strong leadership customer focus, a motivated workforce, malleability and other critical circumstances. Design for Affordability permits an individual to start with a solid foundation or build a solid foundation for products and/or services based on Customer requirements, an effective strategy, a formidable structure and efficient systems for delivering Value.

Here are some key factors that are commonly mentioned that contribute to business success:

- Perseverance (Note: The #1 Cited factor of all successful businesspeople and businesses.)
- Clear Direction: Purpose, Vision, Mission, Strategy, Product and/or Service
- Strong Leadership and Stable Management
- Customer Centered
- Motivated People
- Innovation and Adaptability
- Sound Financial Management
- Effective Marketing (Note: Listed last because this is not always the case)

***What is not on the traditional list of success factors, that has shown to be a strong factor of success, is Design for Affordability.

Design for Affordability has three facets; 1. Customer Value, 2. Creation and Delivery of Value, 3. Cost associated with Customer and Value Elements. It all starts with knowing Customer Value factors intimately (i.e., requirements, needs, wants, wishes, expectations and the amount of money the customer is willing to spend on value). The Value created, must be delivered in a manner acceptable to customer at an expense that keep the organization profitable. Cost for the Customer (i.e., the price the customer can afford, and is willing to pay) and for the Value provider (i.e., the expense and spend on all resources for creation and delivery), must be balanced in a manner that permits profitability and prosperity. The overall Aim is to balance Value, Customer and Cost such that the organization succeeds.

DOI: 10.1201/9781003772934-6

A SOLID FOUNDATION

It would be nice to start 'right' the first time. Too often, in fact, most of the time, companies are started without enough, or the proper amount of preparation. In my research and experience, I've discovered several reasons start-ups occur, beginning with a brilliant idea or invention, to fulfillment of a customer need, to an application of a technology, to a product or service for a specific customer, or by filling of a gap in a market, are all my clients started their businesses with some level of preparation, but for the most part, not an extensive degree of groundwork. Research and technology centers, business incubators, angel and venture capital firms, are all in place to assist entrepreneurs and business founders. I've seen successes using such resources, and I've also seen success stories, like my own 30-year story, happen without that type of help or assistance, other than the use of advisors, guides, coaches and mentors. The best solid foundation appears when perseverance is combined with value, customer consumption and fiduciary soundness. In other words, a foundation based on Affordability.

Value, Customer and Cost are the leading variables for establishing a foundation solution. Value, defined by customer requirements, from creation to delivery, establishes the Value Stream. Customers who can, and are willing to pay the price, established from customer research, provides the critical component in the order-to-cash cycle for revenue. Customer research and market familiarity determines consumer targets. Finally, the details of cost, from value creation and delivery expenses to customer price and revenue generation, combined with income and expenditure proformas, completes the Affordability foundation for operation of the enterprise.

On the human side of establishment, the core principles, values, standards, purpose and culture must be conveyed and established at the outset. The human factors can only be realized if leadership aligns and behaves in concert with such shared ethics and beliefs. Regardless of the stated desires of leadership, a culture will be formed though daily interactions and everyday operations. Norms are ascertained through repeated human-to-human encounters.

For any organization to realize its success, or status of performance, aims, targets, goals and objectives must be present, measured and communicated. Whether it be tangible scoreboards or digital displays, the people need to know if they're winning or losing. In addition, the people need to know what to focus on to achieve success. Such metrics should be both quantitative and qualitative, objective and subjective, realistic and achievable.

Finally, leadership (for the people dynamic) and management (for operational and process dimensions), must be clear and well instituted. Leadership is for leading the people. Management is for administering the enterprise. Both factors drive the 'engine' that powers the systems of success.

When all is ready, it's time to prepare for designing products, services, systems and processes for Affordability.

STRATEGY

Design a strategy such that the value you deliver, and the method by which you deliver it, provides a unique experience for the customer. Develop that strategy such that the price is competitive given the value of the product and/or service. From the strategy, create a plan of action that is focused on the aim. Within that aim, establish targets, milestones and objectives. If necessary, go back to the strategy chapter and review the necessary elements for an Affordability Strategy.

STRUCTURE

Upfront and foremost, for Affordability, Structure the organization for execution of the Value Stream. The traditional departmentalized approach is not only antiquated, but ineffective for focusing on the delivery of value. For the most part, departments are internally concentrated on their own work. Conflicts with other organizations arise due to the inconsistency and conflict between other department's priorities and one's own departmental priorities. Concentration on Customer and Value Stream permits priorities to be set across the organization. Create a structure that aligns with the Value Stream. Below is the summary view of the structural view of the Value Stream;

Suppliers →	***The Value Stream***	**→ Customer**
	Value Creation and *Delivery*	
[	Value Stream	]
[	Value Stream Support Systems	]
[	Business Support Systems	]
[	Leadership Support	]

The primary layer of the structure is the Value Creation and Value Delivery tier. The tier below contains the resources that directly support the Value Stream. The third tier of the organization is populated and tooled with the resources that support the business (e.g., financial, human resources, information technology, maintenance, etc.). Leadership is shown here as the bottom layer which is responsible and accountable for all of the enterprise's support.

At every layer in the structure, system performance must be measured. Historically, individual performance reviews through traditional annual evaluations have been the modus operandi for most departments, managed by the HR section of the company. Affordability is determined by enterprise and process performance, where optional individual assessments can be used. The focus on performance is how well systems and processes serve the value stream. Performance is rewarded through system success.

- System and Process Performance Metrics and Measures. All levels, from systems to processes to procedures to team/individual echelons must be established. Both qualitative and quantitative measures should be considered. For the human side, satisfaction and delight, which is subjective, should be base lined, monitored and displayed. For the objective metrics time, quality and cost should be established. All of these metrics and measures should have some linkage throughout the system

- Affordability Scorecard for the Overarching Enterprise. The suggested 'balanced scorecard' for Affordability contains Customer, Business, Process, People. This is quite similar to the original Balanced Scorecard defined by Kaplin and Norton in the mid-1990's.
  - Customer. Customer, beyond satisfaction, may include on-time delivery to request, meet/exceed customer requirements, quality and reliability from the customers' perspective, value and cost, etc.
  - Business. Of course, revenue, margin and profitability should be measured. Market size, market share, segment, section, quantity and penetration can be measured. Cost and expense reduction, efficiency and effectiveness of investments and spending, order to cash cycle, and growth are all potential metrics to monitor.
  - Process. Process and procedure metrics and measure are dependent on the process and the type of value being delivered. Speed, cycle times, lead time, uptime/downtime, quality, quantity, cost and improvement are all candidates to monitor, display and utilize for managing and improving the system elements.
  - People. From human resource requirements to knowledge and capability, to skills and abilities, to efficiency and effectiveness, to growth, for achievement and success are the possibilities for measurement of the human resources.

"How do you know if you're winning or losing if you don't know the score?"

SYSTEMS

Affordability defines systems as a composition of integrated processes comprised of procedures and tasks that all support the value stream. Whether the systems create and deliver the value, or the systems support the entire realm of the value stream from suppliers through the value stream to customers, all functions are focused on supporting suppliers, creating and delivering value, and supporting customers.

System components are primarily integrated processes that are constructed of procedures and accompanying tasks. This linkage ties in all individuals, machines, tools, materials, methods, means, monies, management and value competencies to the purpose and performance. The hierarchy of enterprise, systems, processes, procedures, tasks and people is in alignment with the structure of strategy, operations, tactics and individuals.

Systems assist in defining the Enterprise performance through strategic use of the operational and tactical entities. Affordability Systems provide performance status through the use and utilization of critical performance measures of: Customer, Business, Process, People. Each system has metrics that tie back to a Balanced Scorecard of Affordability. The Aim of Affordability is to provide systems that keep these metrics and measures are in alignment with the Delivery of Value with the transaction of Order-to-Cash Customer purchase. The monetary or fiduciary focal point is not the utmost measure of success. Affordability happens when the Customer, Business, Process, People measures are balanced and equalized throughout all the enterprise's systems.

What not to do regarding addressing systems: *"Learnings from a Healthcare Clinic project in Cleveland Tennessee."* When an Enterprise is focused on primarily money or money alone, the entity is off balance. Without consideration of Customers, Processes and People, using the money metric of the Business portion of the Enterprise limits the view of the success perspective. A few years ago, I was invited in to a clinic of a nation-wide Healthcare Corporation to provide input as to how to increase performance and profitability. On my first visit, I spent four days touring the facility, observing the processes, documenting the value stream and executing a few Kaizens and 5S events, for creating some small successes and 'harvesting' some 'low hanging fruit'. On the second visit, I followed up with several physicians and nurses to map and record their standard value stream (i.e., examining and treating patients.), and to perform a few additional Kaizens and 5S events. The results of the two visits provided a good rendering of their primary value stream map, documentation of their value stream processes, calculate process cycle times and gather employee satisfaction measures as to the results of the Kaizens and 5S results. In Summary, the Value Add Cycle Time Measure of physician exams of patients came to ~20 min (Note: with a range of 15–30 mins), which amounted to an average of 15 patients examined each day for each physician, with a measurement of Employee Satisfaction increased from a baseline of 86% to over 90% as a result of the improvement activities, and a brief calculation of waste reduced from Kaizens and 5Ss (mostly from reducing over-inventory spending and materials wasted from expiration) amounted to more than $100,000.00 in a year. After spending the two four day weeks with the client to assess their current state, and execute a few 'quick wins', I had a brief one-hour meeting with their CFO for the purpose of providing an update and suggesting some activities and events for improvement. I showed him the results of the two weeks of engagement and explained a short-term strategy and schedule I planned to execute. His response was a bit perplexing and confusing; "This is all great and good, but I have an established goal of 22 patients per physician each day to generate the amount of revenue I want to increase profitability" (Note: That meant, *every physician* would have to spend less than 15 min with a patient on every patient visit. This set an alarm off in my head … he is jeopardizing patient safety, care quality and sacrificing physician integrity for revenue.). In open honesty, since the CFO was leading this effort, I had to walk away from the project and find reasons not to go back (Note: This was a rare incident for me that only happened twice in over 300 projects). Lesson Learned: When only one aspect of a system's success is measured (i.e., Money), without considering all dimensions (i.e., Customer, Business, Process, People), this imbalance most often results in failure. A Realized Fact: Within a few years, physicians left, the clinic was closed and the facility sold to another competitive healthcare provider in the area.

DESIGNING SYSTEMS FOR AN ENTERPRISE

For Affordability, the systems for the Enterprise either define the Value Stream in that they are dedicated to creating, producing and delivering Value to the customer, or support the operation and execution of the Value Stream (e.g., procurement, facilities, logistics, sales, marketing, customer support, finance, legal, human resources, technology, training and documentation, etc.). The systems and processes either serve an external customer, or serve an internal customer, or both. It is also the case that the systems and processes will have suppliers and support functions for the function they execute. In classical systems thinking, they form a S.I.P.O.C. (i.e., "S" = Suppliers, "I" = Input, "P" = Process, "O" = Output, "C" = Customer). Suppliers provide the Input to the Processes in the Systems, and the Processes provide Output for the Customer (i.e., Internal or External). The design of systems includes all processes, their suppliers and inputs, and their outputs and customers. A system is the integration of all processes in the System: Suppliers → Inputs → Processes → Outputs → Customers.

System Processes can be broken down into sub-processes and micro-processes or can be described as an integration of procedures which are made up of tasks transforming inputs to be valuable outputs for "customer(s)" of the process. Processes and procedures use and utilize manpower, materials, machines, measurement, methods, tools, tasks and other to accomplish their defined function and purpose. In designing Systems and Processes attention to detail is of utmost importance. Enterprise systems and processes form the Value Stream and serve to define the Value Stream Support functions that deliver Customer Value.

DESIGNING VALUE FOR AFFORDABLE PRODUCTS AND SERVICES

Value is what is created, produced and delivered to the customer as an output of the Enterprise. Customer requirements are the prerequisite for Value Design. The Customer's wants, needs and expectations are at the center of every Affordability Design. Since Customer requirements define Value, it only makes sense to design products and services to meet and exceed customer requirements. Often, step one is to design form and function, meeting requirements and purpose for the customer (NOTE: Considering all types of customers, every applicable market and existing competition.).

At times, as requirements change, products and services need to be improved and/or 'adjusted' to meet new and changing customer demands. Newer technologies, new features, additional requirements and expired requirements all need to be addressed. The incorporation of new functions or the termination of expired features is an ongoing necessity in every product or service life cycle.

While I served as a product manager for NCR in the mid-1980's, I was first exposed to the theory and philosophy of delivering Value through product functions and features that meet customer requirements. Although at that time, I had not synthesized the Theory of Affordability nor understood the Philosophy of Value Delivery, I was coached and mentored by my director to create an inventory of functions and features of my product that met customer requirements, missed customer requirements and did not meet nor offer specific customer requirements. This forced me to research and create a comprehensive customer requirements checklist for the customers in my product's target market. At the time, since my predecessor did not provide a customer requirements list, it was a laborious task to create one. Today there are numerous tools on the market for this task, however, nothing beats the old fashion method of personal effort and attention to detail

(Note: I suggest using a team of five to eight professionals, to create, manage and maintain the list. For an individual, it is an arduous task, and "two heads are better than one" when it comes to verifying, validating and populating the requirements list. Otherwise, pay a professional research entity to deliver the feature and function inventory.).

The next step is to put your product or service design team to work to shape and form the function and features into something that can be understood and delivered (Note: The design team may be only one person … you.). Images and illustrations work best to convey the product or service vision with any additional detailed drawings, graphics, artwork, diagrams, sketches, applicable visuals and visual aids that communicate the model and concept to be produced. The more 'understandable' your concepts and constructs are to others, the easier it will be to achieve 'the perfect result' (Note: Facts, data sheets and descriptions are also helpful to support your visual depictions and renderings.). The next step should focus on details and decisions for creation, production and delivery.

With the customer requirements list in hand, and a rendering of an image or illustration of the product and service, the next step is to determine what is needed to create, produce and deliver the Value, and what needs to be done to ultimately accomplish the delivery of Customer Value. Product and Service design and planning, and process design and planning are all needed to complete the necessary Value Delivery Systems. My recommended toolset for defining and deploying the mandatory components of Value Stream are the House of Quality (HOQ) and Quality Functional Deployment (QFD). At this point, it's necessary to mention, I have witnessed the use of these tools only 3 times in the past 35 years since I've been exposed to them (Note: Twice using the originally intended approach and once using an alternate scheme of application.). Although I recommend this method to every organization I've served (i.e., 137 in 35 years), I've used it once myself, and it's been used twice by clients. It seems to come across as too difficult and too time consuming to utilize (Note: Which, honestly it is difficult and time consuming, but it turns out to be ultimately beneficial, providing comprehensive detailed information for use when designing and delivering a product and/or service.).

Some friendly advice, before beginning using the House of Quality and Quality Functional Deployment: Seek to fully understand the tools before you start. There are several books and workshops available, and you should practice and experiment before 'jumping in'. One of my favorite sayings, "Always try, pilot and verify before going '*Customer Live*'."

HOUSE OF QUALITY (HOQ) AND QUALITY FUNCTIONAL DEPLOYMENT (QFD)

Once the customer requirements are known, it's time to decide the details of the product and service design and the processes and process parameters necessary to deliver the Value to the Customer. Using the House of Quality (HOQ) and Quality Functional Deployment (QFD) the components of the product or service become well known and the detail for the mandatory processes are determined.

In order to get started, form a team of five to eight knowledgeable individuals that are intimately familiar with the product or service being studied, assessed or evaluated. Get the team to spend some time creating a list of Customer requirements. Be sure and include all the needs, wishes, wants and 'dreams' customers may have (Note: This list will be considered the "What's" of the HOQ.). In addition, have the team list and identify all the design requirements and methods the Value Proposition uses to meet the Customer requirements (Note: This list will be considered the "How's" of the HOQ.). Now, with both lists in mind, the team is ready to begin using the House of Quality and the Quality Function Deployment methodology (Figure 6.1).

HOQ is a Two-Dimensional Matrix of Thinking. It basically compares the "What's" (i.e., Customer Requirements) with the "How's" (i.e., Design Requirements or Methods of meeting the Customer Requirements).

HOQ provides a cross reference of any two items of the "What's" and the "How's" identifying the relationship between them (i.e., "the body"), while illustrating the correlation between each method (i.e., "the roof"), displaying the 'amount' or degree level of each method (i.e., "the bottom"), and the competitive assessment of each requirement with others in the market or industry (i.e., "the right side").

QFD was originally developed in Japan in the 1960's for product development and quality management. The primary goal is to ensure that Customer Requirements (AKA: "Voice of the Customer") are fully integrated into product and service design,

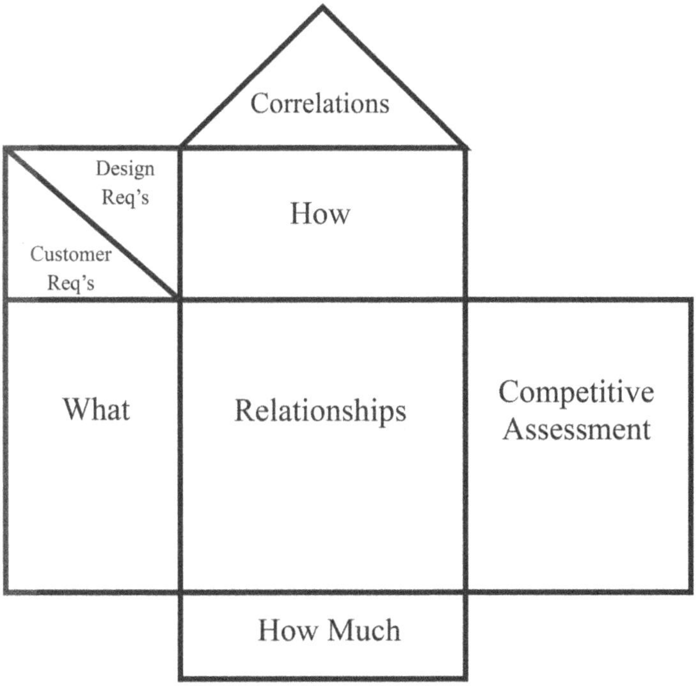

FIGURE 6.1 The house of quality model.

and Value Stream Processes. The key components of QFD are HOQ, Customer Needs and Technical Requirements. The aim of QFD is to meet and exceed Customer Requirements to achieve Customer Satisfaction and Delight, reduce development time and time to market, and enhance communication throughout all team and entities involved.

As QFD evolved and improved throughout the 60's, the House of Quality (HOQ) appeared and was first used by Mitsubishi in 1972 at their Kobe shipyard. Soon after, HOQ was further developed by Toyota for internal use, it was deployed to Toyota Suppliers for their utilization. HOQ is effective because it provides a visual map of the relationships between customer requirements, design requirements and process requirements. Additional features involve design correlations, design objectives and measures, and competitive analysis (Note: If the user chooses to evaluate the competitive dimension). Beyond the first dimension, there are three additional dimensions that delve into the relationships of design requirements with parts and part characteristics, parts and part characteristics as they relate to critical process parameters, and finally how critical process parameters relate to process control characteristics (Figure 6.2).

The full use of the HOQ and QFD involves a four-phase approach that begins with customer requirements, and ends with process parameters and process control characteristics:

I Product/Service Planning → II Design → III Process Planning →
IV Process Control Planning

Initially, in Phase I, Customer Requirements (AKA: Customer Desires, the "WHAT"s) are entered into the template, along with Design Requirements (i.e., These are the "HOW" items that are planned to be used to address the "WHAT" elements. Design Requirements are how you intend to 'tackle' each "Customer Desire" or Customer Requirement.). The "roof" is used to convey the correlation between two design requirements. The bottom (or 'basement') contains the how much, quantity

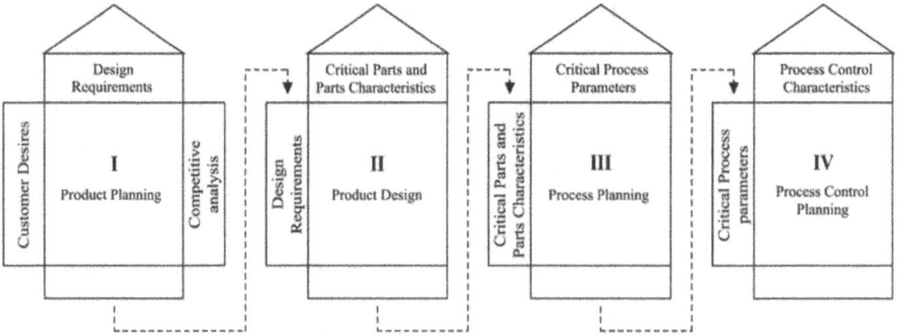

Source: Cohen (1995)

FIGURE 6.2 The four phases of the house of quality.

or amount of each design requirement. The right side of the matrix is competitive benchmarking for displaying how the competition rates on each customer requirement. This first phase provides a team with a visual for discussing what is required, how much is required, how design requirements correlate to each other and how well the proposed or current solution compares to the competition. This permits a team to make decisions on changes or adjustments needed to stay competitive and be the best as compared to the competition. It also allows the team to view, discuss and improve the Value being Delivered to the Customer.

Phase I set the stage for deciding what to do to deliver the Value to the customer from the machines and methods to the processes and procedures. It permits the QFD Team to research, discuss and decide what approach is best to employ and how the functions will be deployed (Figure 6.3).

Below is a simple sample the ASQ provides for Phase I of the HOQ.

As you can see, the relationships between Customer Requirements and Design Requirements are 'scored' (i.e., 'scored' by the individual or evaluation team building the HOQ). The 'roof' communicates the correlation between two Design Requirements. Within relationships and correlations, a cell is left blank if a relationship or a correlation does not exist. The bottom shows the risk, the weight and how much is needed for each Design Requirement as well as a technical benchmarking assessment for the competition. The right side of the matrix presents how the

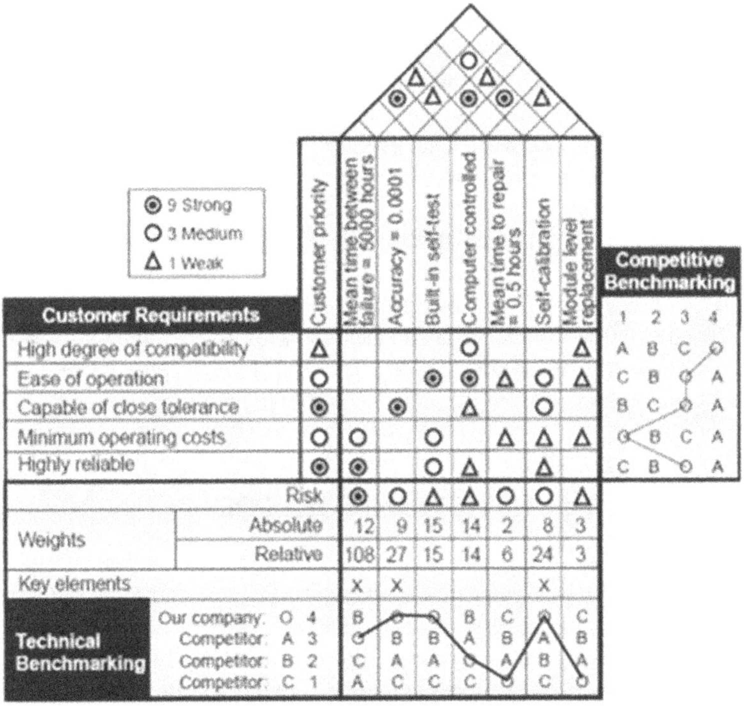

FIGURE 6.3 A basic example of the use of the house of quality.

competition performs on each Customer Requirement. HOQ Phase I just establishes the basic criteria for planning a product or a service, and it attempts to assist an individual or design team 'see the product' or 'see the service', and make decisions regarding how to satisfy all of the Customer Requirements. This example is simple and straight forward, a real implementation is much more complex and might take an entire 'neighborhood' of HOQs to address an entire product or service.

HOQ Phase II is focused on product or service design. The Design Requirements are now the 'WHAT's and the critical parts or component characteristics are the 'HOW's of this phase (NOTE: Using an Excel Template, the 'How's of Phase I or spreadsheet 1 are automatically populated as the 'WHAT's of the Phase II spreadsheet). Now, the individual designer or design team can assess the relationships and correlations of the characteristics and determine how much of each characteristic will be required.). A competitive analysis is not required after Phase I since specific design requirements and process requirements are unique to the product or service being designed.

HOQ Phase III focuses on process planning. Knowing the detailed characteristics for the design requirements (i.e., now the left side of the matrix) and the critical process parameters (i.e., the 'HOW'), the relationships, correlations and how much's can be inserted, discussed, scored and determined. At this point, the design individual or design team can make process specific decisions for defining the necessary processes and procedures required to create and produce the product or service.

HOQ Phase IV takes the critical process parameters and views their relationships with the critical process control characteristics. The correlations and how much's are assessed and evaluated to complete the HOQ event. This is not a single hour activity, and it often takes days, if not weeks to complete. When it is finished, it provides a comprehensive visual for how to meet customer requirements with a comprehensive illustration of what goes in to make the product or service, how it will be produced, the processes that will be used, and the extent to which the processes will operate.

As the system is put in place, don't be surprised if after piloting, validating and assessing the processes and final output, changes are needed to tune and adjust the methods and procedures. In the few instances I've had with HOQ, the detailed work upfront yielded a faster preparation and execution for time-to-market. Faster-Better-more Affordable was the result. On one occasion, I was exposed to an alternative approach and method of using the HOQ and QFD to address Quality of Life.

A UNIQUE USE OF THE HOQ AND QFD FOR HEALTHCARE QUALITY

Several years ago, I was partnered with the IISE serving the Georgia Hospitals Associations (GHA) initiative to train, develop and certify healthcare professionals across the state of Georgia for systems improvement using Lean Six Sigma means and methods. We were able to certify hundreds of Green Belts and almost 100 Black Belts via a program accredited by the IACET. Each Black Belt was required to pass an exam and complete an approved project in their facility for improvement of an important process in the areas of effectiveness, efficiency, speed, quality, cost and other critical performance measures. One project in particular made use of HOQ and QFD for "Reducing Variances with Total Knee Arthroplasty (TKA) and Total Hip Arthroplasty (THA) Surgical Procedures that Lead to Risk for Complication." Certified Lean Six Sigma Freya Gilbert RN used the project to earn her certification (Note: I used this project as an example of improving healthcare in my last book "Healthcare Affordability." For additional project details you may want to reference that publication.).

After THA and TKA surgery, complications may occur that impacts quality of care for the patient. As part of a patient safety initiative Freya designed her project to use the HOQ methodology for identification of possible preemptive treatments and counter measures to ensure quality and safety. The overarching goal was to standardize processes using multidisciplinary teams through identifying effective standards for processes and procedures. As part of that goal, HOQ was used to study common complications related to THA/TKA and identify strategies for mitigation. Objectives include: "Voice of the Customer" (Listen to the Patient), Patient/Provider Partnership, Improve Patient Outcomes by reducing—Surgical Site Infections, Returns to Surgery, Readmissions, Length of Stay, Costs.

The multidisciplinary team utilized QFD to analyze quantitative and qualitative data to assist in the formulating standardized, evidence-based practice that considers patients' needs and preferences. HOQ was used as a relationship and correlation template to evaluate customer and technical requirements. Other tools used with this project were Failure Modes Effects Analysis (FMEA) and Plan-Do-Check-Act (PDCA).

Instead of using HOQ in the traditional manner, Freya and the team used the tool to dig into the "Demanded Quality" (Customer Requirements or Patient Quality Procedure Steps: Procedure Scheduled, Pre-Admission Testing, Preop Holding, Intraoperative Period, Post Anesthesia Care Unit, Inpatient, Discharge) as related to Quality Characteristic Areas (Clinical, Equipment, Education). This effort took several sessions before all inputs were provided and the HOQ was ready for discussion and scoring. The team then scored the relationships and correlations. The Target or Limit Values were inserted and the rankings were applied. Note that the competitive analysis was not needed for this particular application since the team research and study was for application in their facility alone and not related to commercial competition with other healthcare organizations. After several weeks of team working session, Phase I of the HOQ was considered complete (Figure 6.4).

This was the result of Phase I (Credit: QFD Online was used at www.qfdonline.com).

This view provided the team with insight into the strength of Customer Requirements ("The What's") as compared to the Clinical, Equipment and Education Characteristics ("The How's"). The team was able to now understand some detail regarding how clinical, equipment and education characteristics correlated, the amounts ("How Much's") and the priorities using ranking and weighting of each element. After Phase I was discussed and analyzed, the team was on to Phase II. Just as it was for Phase I, it required several weeks of development and preparation for Phase II to be complete and ready for discussion, analysis and planning (Figure 6.5).

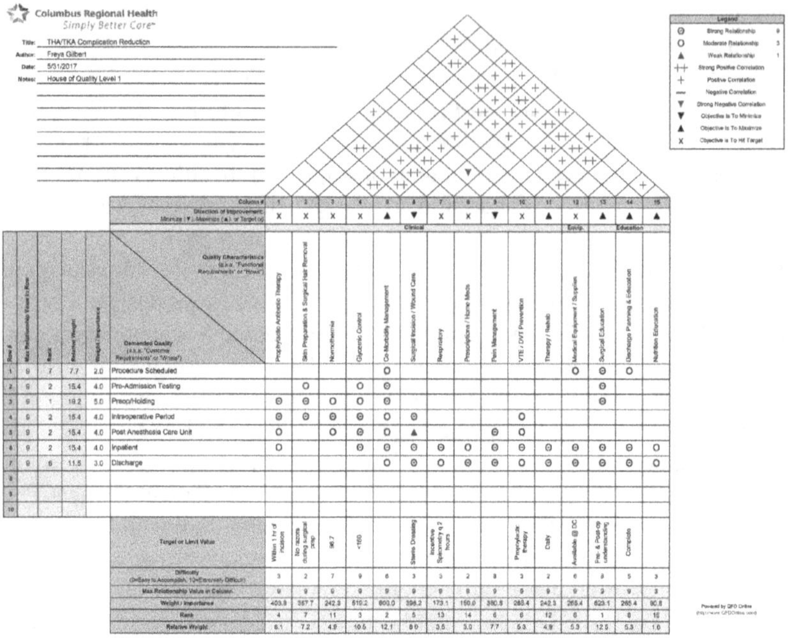

FIGURE 6.4 House of quality level 1—THA/TKA procedures and clinical/equipment/education.

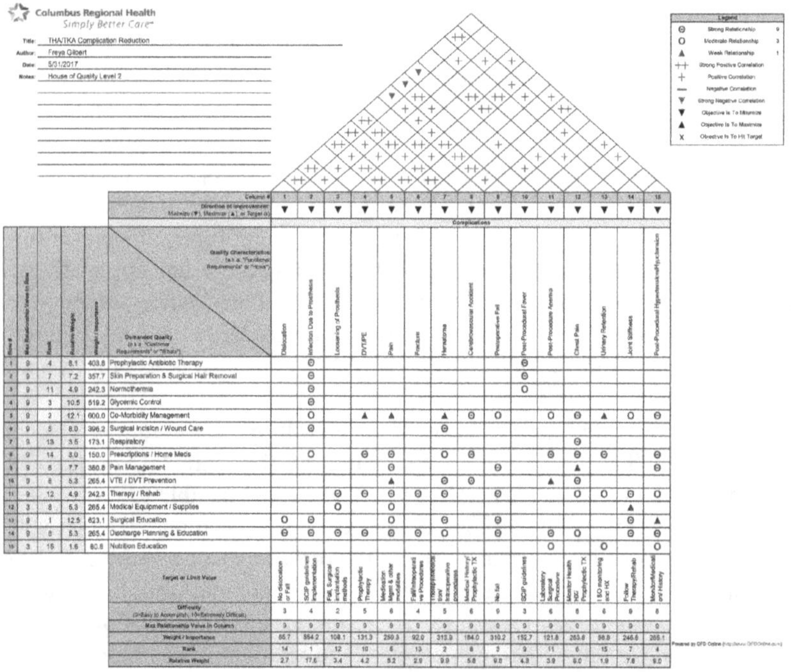

FIGURE 6.5 House of quality level 2—clinical/equipment/education and complications.

The team concentrated on populating the data for Phase II (The Quality Characteristic Areas—Clinical, Equipment, Education as they related to Complications).

This phase provided the team with a view of how complications related to the quality characteristics and how the complications correlated with each other. This level of understanding created new insights related to possible applications, interventions and preemptive activities. By this time, all members of the team began to appreciate and recognize the power of the HOQ tool. In parallel with the team's work of populating the matrix, ongoing research, study and discussion with patients, physicians, nurses, administrators, technicians and all related resources (Figure 6.6).

Following Phase II, the next step was to complete Phase III. This Phase illustrated the relationships between complications and how they related to comorbidities and financial factors.

The comorbidities and financial characteristics were taken from actual records of 12 months of in-patient medical history. Using complications and their relationship to comorbidities and financial factors, combined with the correlations and target/limit values, the team was able to attain additional intelligence for diagnosing problems and challenges facing physicians, nurses, techs and the facility. At this level of research, significant discoveries were realized. Also to note, the interest level grew,

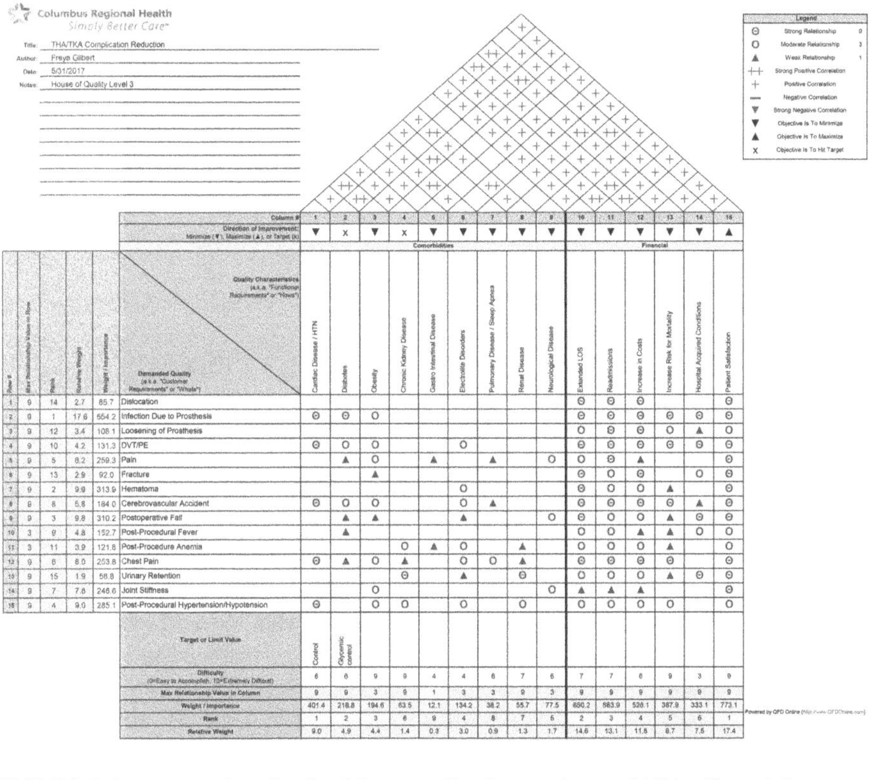

FIGURE 6.6 House of quality level 3—complications and comorbidities/financials.

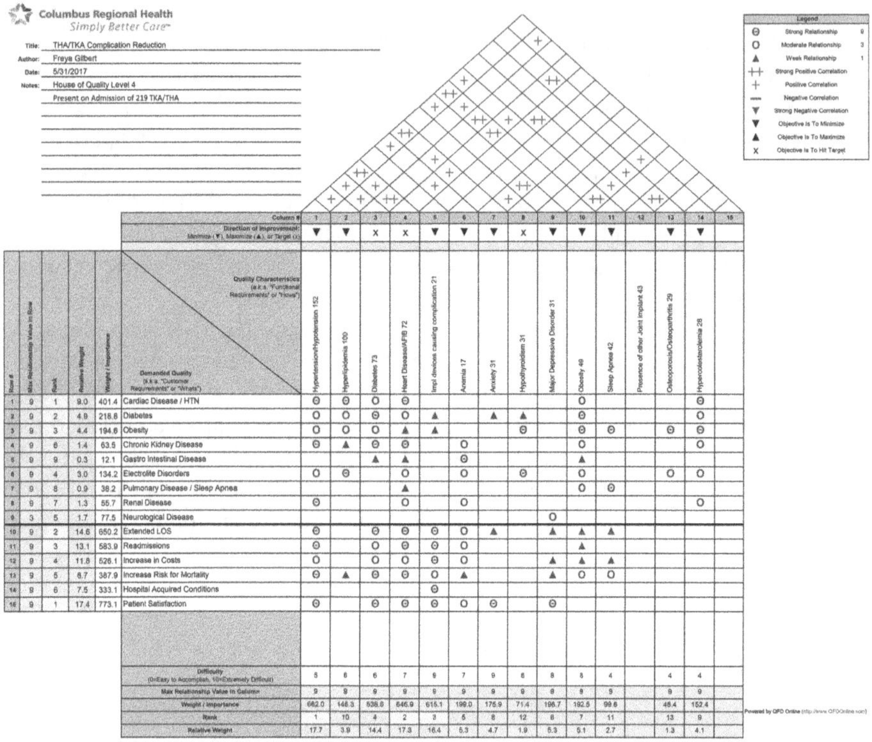

FIGURE 6.7 House of quality level 4—comorbidities/financials and conditions on admission.

the staff was more enthusiastic about participation and the attention of the leadership increased (Figure 6.7).

Following Phase III was Phase IV that involved comorbidities and financials compared to related conditions:

This shows actual diagnosis coded post operatively on patients. It opened the team's eyes to some diagnosis' that needed more attention such as Obesity and Sleep Apnea, etc. The next steps were to drill down into occurrence reports for opportunities that may have arisen with depression, hypertension, falls and infections. The team was now able to dig down and explore how to take care of each area. Now it was crystal clear that work to standardize care was absolutely necessary to achieve "best practice" across the board (Figure 6.8).

And, on this rare occasion, there was a need for Phase V.

At this level, the team wanted to review and look at what was current and in place, or was required to be implemented. This area was also leveraged to review occurrence report history.

Although it took a good amount of time and effort by the team, the results and the accomplishments were quite beneficial. At each level of the HOQ, specific topics

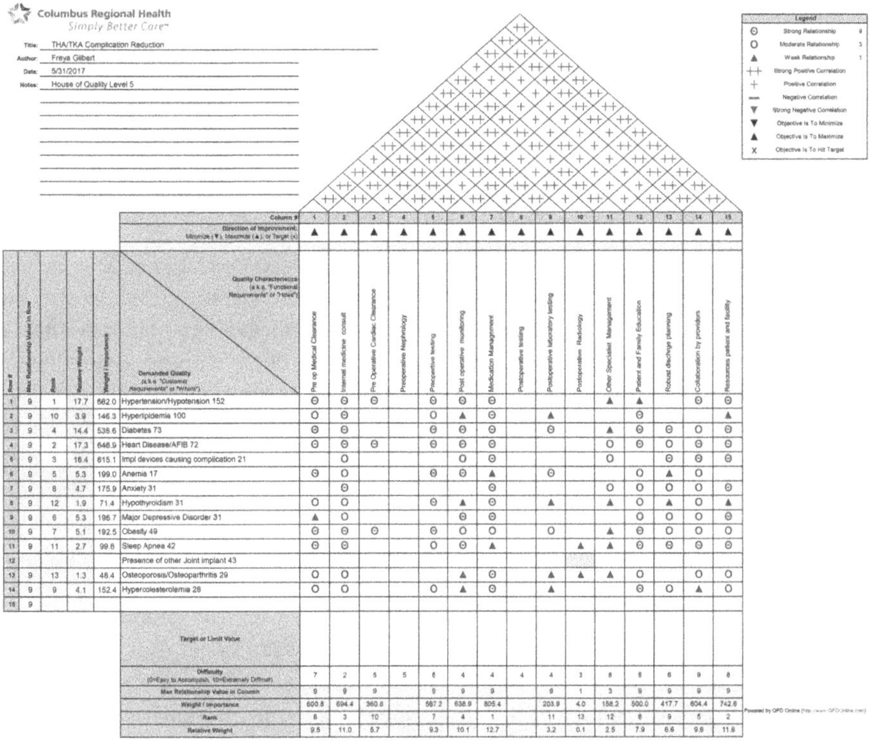

FIGURE 6.8 House of quality level 5—conditions on admission and treatment.

came to the forefront for the team to discuss and prioritize. The prioritized topics were then tagged for action.

- Phase I: Surgical Education, Comorbidity Management, Glycemic Control
- Phase II: Infection due to Prosthesis, Hematoma, Post-Operative Fall
- Phase III: Cardiac Disease/HTN, Diabetes, Obesity
- Phase IV: Patient Satisfaction, Extended Length of Stay, Re-Admissions
- Phase V: Medication Mgmt, Patient & Facility Resources, Internal Medicine Consult

As a result, the team chose to focus and work on: falls, reduction of risk of surgical site infections, increased medication compliance (especially in risk populations with additional comorbidities).

The primary finding was stated as follows: "The development of standardized clinical care pathways to ensure consistency of care throughout the continuum". This includes the use and utilization of Physician Champions and Various Specialties (e.g., Orthopedics, Medicine, Anesthesia, Rehabilitation) to "walk the current processes" to ensure and assure they are stable and predictable (Including, but not limited to: Antibiotic Therapy, Standardization of Order Sets, Skin Preparation, Medication

Therapy, Skin Closure Techniques, Methods and Time Frames for Drainage Devices, Methods and Supplies for Incisional Dressings, Education for Patients and Care Givers prior to Procedure, etc.).

Learnings of this project were stated as follows: Senior Leadership Support is Mandatory, Physician Support and Involvement is Required, In-depth Research and Data Analysis is Valuable, Selection of the Right Team Members is Critical, Importance of Prioritization of Project Initiatives is Foremost, and the practice of "Do Not Make Assumptions!" is Essential.

Resulting Actions: Develop and Implement Standardized Patient Education in various formats, Patient Optimization Methods instead of Preoperative Clearance, Focus on Prevention, Partner with Patient. Some focal points include: mailing education to patient, having education available in the physician's office, mailing sage wipes to patient, teaching patient what they will need at home, conveying what type of assistance they will need at home, developing a physician documentation tool that reviews specifics and will let the surgeon know if the patient has chronic disease with their current state (e.g., HA1C, Sleep Apnea needs, Renal status, Chronic UTI, Chronic Prostatitis, Dental Infections, Asthma, Depression/PTSD, Nutritional status), patients responsibilities such as prevention of safety risks at home (e.g., falls, dietary, proper wound care).

Although, in this case, the application of HOQ was non-traditional, as prescribed, this creative approach to QFD accomplished a great achievement in improving patient care and quality of life for patients in this system. For additional uses and application of HOQ and QFD I suggest that you research, read and educate yourself on this powerful approach to identifying and designing Affordability Systems and Processes.

PRODUCTION PREPARATION PROCESS: 3P

Another powerful tool used in the Affordability Toolbox is 3P (Note: 3P is a tool that emerged from within Lean Manufacturing applications. 3P is a three-word acronym consisting of the words "Production Preparation Process." It is a methodology designed to create efficient waste-free systems by focusing on concentrated, thorough preparation and process design. It is a proactive approach used to design and develop new processes, new products or new services using a structured approach. It is also used to re-define a system that is in place and in need of re-design. Production is the means by which Value is created, produced and delivered. Preparation involves setting up the necessary environment for successful production (Selecting the right resources, right tools and machines and right procedures for a smooth flow of the Value Stream). Process is about designing efficient and standardized processes for production of Value. The ultimate benefits include elimination waste, decreased cost, improved quality, reduced lead time and enhanced employee engagement. The outcomes and accomplishments have proven to be greater customer satisfaction and better system performance.

3P FOR THE UNSINKABLE

About six years ago, I was invited in to deliver a 3P program for Boston Whaler in Edgewater Florida. At that location, Boston Whaler produced 4 of their boat models: Super Sport, Dauntless, Montauk, Outrage. The orders and demand for Super Sports,

Dauntless' and Montauks outpaced their production capacity. Boston Whaler was looking at an alternative site to produce the additional demand using a vacant facility they owned in Merritt Island Florida. A little history before we dig in ... The first Boston Whaler Boat was invented by Richard "Dick" Fisher (a Harvard Graduate) in 1958 in his garage in Braintree Massachusetts. He introduced the concept of using foam filled hulls which virtually rendered the boat "unsinkable" (i.e., it is known, "you can cut them in half and both halves will continue to float"). This innovation was a gamechanger in the industry and set a new standard for safety. Today Boston Whaler is produced under the parent company of the Brunswick Boat Group (Boat Types Produced: Bayliner, ***Boston Whaler***, Brunswick Commercial and Government Products, Crestliner, Cypress Cay, Freedom Boat Club Harris Pontoon, Lowe Boats, Lund, Meridian Yachts, Princecraft, Quicksilver, Rayglass, Sea Ray, Uttern, Valian).

In order to prepare for the event, I visited the Edgewater facility and emersed myself in their current processes in order to gain an in-depth understanding of the systems, processes and procedures being used. While I was there, I asked the General Manager, "Why don't you just replicate the successful production lines today in the new facility and keep production consistent at both sites," His response, "We're bringing in some expert resources from around Brunswick for this event so we can design new systems that can be implemented at other Brunswick subsidiaries beyond just Boston Whaler." After completely understanding the customer's requirements (VOC, Voice of the Customer) and studying the assembly processes for the targeted three models, I returned to my office to prepare a 5-Day 3P Workshop for the people attending the event.

The session was organized and executed by a very young, very talented, very capable and very eager Industrial Engineer who graduated from The University of Central Florida which is located about an hour away from Edgewater and also about an hour from Merritt Island (the target site). He and I formed a strong cooperative and collaborative bond which was a significant advantage for me since he knew both sites intimately (Note: He was recruited by both Boston Whaler of Edgewater and NASA on Merritt Island). Since 3P uses a hands-on, learn by doing, experiential learning approach for design, a lot of materials are required, some that have to be sourced within an hour of need. It took us several days to design a solution and prepare a plan.

Edgewater Florida being the flagship site for Boston Whaler was also a lead influencer of boat production practices for The Brunswick Boat Company. The plan was to spend the first day of the week in Edgewater (Monday), and the last four days of the week (Tuesday–Friday) at the "Old Sea Ray Sailboat Factory" on Merritt Island where the new production lines were targeted to operate. About half the attendees were from Edgewater, and the other half were brought in from other locations representing several other Brunswick Boat Company brands. In total, 18 people attended the 3P Event, and all were dedicated to the activities "without interruption" (Note: This endeavor requires full involvement, concentration and engagement of all participants all the time for the entire week. It is a fast moving, fast designing, fast prototyping, rapid planning activity that should not be interrupted in any way ... Group or Individual. This is a mandate and discipline that must be adhered to and policed by top leadership!)

Although Boston Whaler produced four models at the plant, only three were 3P targets (i.e., Super Sport, Dauntless, Montauk). With nine attendees familiar with the site's production, and nine attendees from other location, we predefined three teams of six, with each team having three "locals" and three "out-of-towners." This design provided local expertise as well as potential out of the box thinking from the "out-of-towners." We operated from 8:00AM to 5:00PM every day with a one hour for lunch where we could all "break bread" together and dialogue and discuss topics.

This was our 'Official' P.A.L. (Purpose, Agenda, Limit and Goal) and "Kick Off"

Introduction

- Purpose: Boston Whaler 3P Event
- Agenda:
  - Day 1: 3P Introduction and Information at Edgewater
  - Day 2: 3P Innovation at Merritt Island
  - Day 3: 3P Prototyping and Rapid Design at Merritt Island
  - Day 4: 3P Prototyping and Rapid Design at Merritt Island
  - Day 5: 3P Optimization at Merritt Island
- Limit: 8:00 – 5:00
- Overall Goal
  - 3P Team Developed Prototypes for Evaluation
- Introductions (Name, Area of Expertise, etc.)

When to use 3P

- Design of a New Product using Voice of Customer
- Improving the Design of an Existing Product
- Changing Customer Requirements
- Improvement of Product Quality
- Increasing Plant Capacity to meet Customer Demand
- Relocating Operations or Instituting a New Location
- Making Process Quality Improvements
- Purchasing Capital Equipment

(Note: I'm not going to present all 79 slides in this book, only the pertinent information)

Now that the guidelines, boundaries, reasons, daily activities and usage were framed, and the introductions were presented, it was time to jump in and get the ball rolling. A brief overview of Lean and 3P was provided and the event began (Figure 6.9).

I was able to illustrate the 4 Phase 3P Event on one slide that was used throughout the week to convey where we were, where we are presently and where we would go next:

The primary inputs for Day 1 comprised of "Voice of the Customer" (Customer Requirements), Ideas, Bold Goals (i.e., they didn't think we could fast prototype and define a new system in a week), Research Data. The Four Phases encompassed Information: "Gather Knowledge," Innovation: "Develop Alternatives," Prototyping & Redesign: "Prototype & Converge," Optimization: "Evaluate & Plan." The Steps involved: Product Planning, Product Try-Storming, Converge Designs & Prototypes,

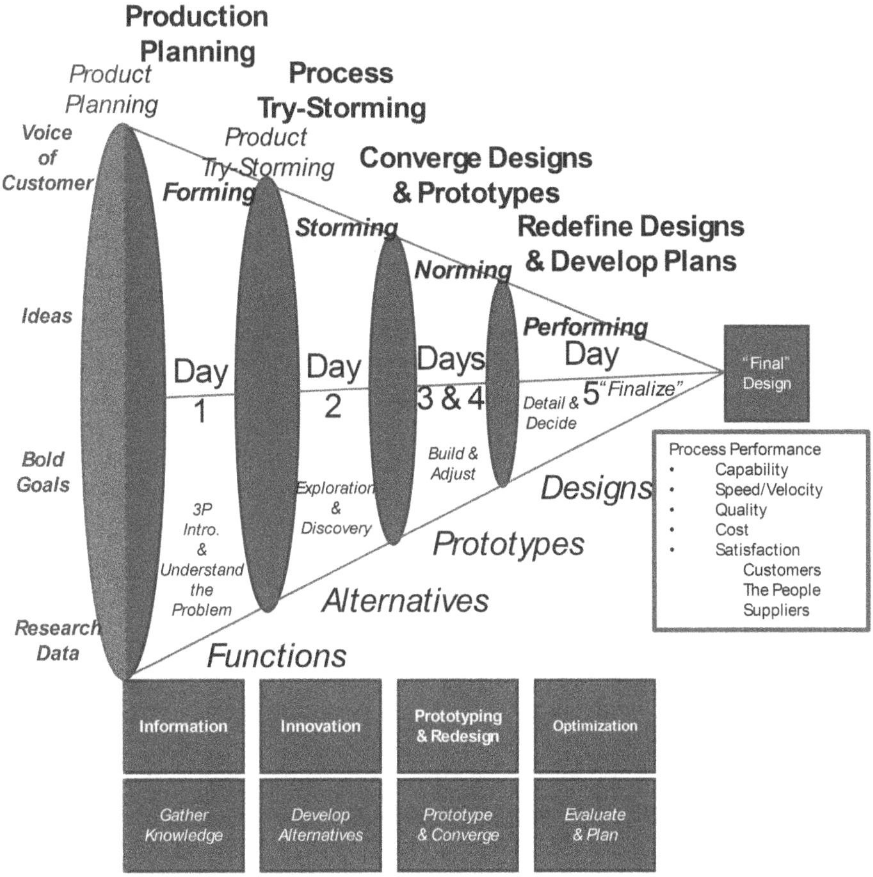

FIGURE 6.9 3P schedule.

Redefine Design & Develop Plans. Functions lead to Alternatives that lead to Prototypes that lead to Designs. Convergence is the Key.

Most of the morning of first day was spent understanding how 3P works, understanding the meaning of the Four Phases (Information, Innovation, Prototyping & Redesign, Optimization), and understanding the roles and responsibilities for each team and every individual. Beginning a couple of hours before lunch and ending mid-way in the afternoon, a Gembe Activity permitted the teams to "go to the floor," observe and record the current state of the systems and processes to be use as reference material Tuesday through Friday. The first day ended with finalization of the Information Stage (Knowledge, Roles and Responsibilities, Tools and Materials Required for Prototyping, Logistics, Goals, Charters and the Disclosure that Top Leadership will serve as Judges to evaluate, assess, pick and identify "Best Ideas and Best Practices" at the end of Friday for convergence into a "Final Design").

The team started Tuesday at the Merritt Island facility with a "walk-about" to get the team familiar with the physical footprint of the factory (Note: Only one person of the 18 participants had ever been there before). We reviewed what we had covered and discuss what would take place over the next four days. The teams divided into their predetermined respective groupings and got right to work. A recap of the morning was held and lunch was served. The afternoon covered value, value add, value stream mapping, flow, root cause analysis, developing alternatives and prototypes, and 'process at a glance'). The day ended with a status and progress check, with readiness check for the next day.

Wednesday began with Process-at-a-Glance, Best Alternatives, Preparation (People, Materials, Tools, Workspace). The update of Process-at-a-Glance in conjunction with the Prototype being developed was accomplished along with assessment and evaluation, including updating and improvement of the Prototype. In addition, the third day also focuses on integration of product functions with Process Prototypes and Alternatives to be validated. The day closed with status reports and progress checks.

The fourth day was dedicated to the physical construction of the prototype. This exercise focused on prototyped processes, physical layouts and scale modeling. This part of the event defined equipment, flow, supermarkets, Poke Yoke, TPM, devices, POU, cell design, work and other details. The culmination was the end of day report out and status check.

The final day required that each team run a simulation of their prototype. Experimentation, piloting and testing with updates, corrections and improvement was most of the morning's focus. It was emphasized to record the event with pictures, videos, visuals and illustrations. The afternoon was reserved for the final prototype design (including costs, resources, timelines, risks, actions, etc.). The leadership arrived at the end of the day, and each team executed a presentation of their final prototype. The leaders provided feedback and selected the best-in-class design elements from each of the three prototypes to be merged and incorporated in a final design. One prototype was not chosen to "win" because all of the prototypes contained procedures that for use that could be incorporated into one superior system.

When we were finished, we held a celebration ceremony for the teams and the participating leaders. With a target prototype in hand, the nine "locals" worked the design for months to "flesh out" defects, enhance the prototype with improvements and ready the completed design for implementation. Since three Boston Whaler Models were the product topics, there were three production systems set up to operate, side-by-side, in parallel within the facility.

The result created three viable prototypes that were merged into a final design. When it was all said and done, knowledge and capability was transferred to 18 Boston Whaler professionals who now, through the hands-on experiential learning exercise, now knew 3P, could design and execute a 3P event, and develop processes and systems with the 3P knowledge and capability for deployment throughout the several Brunswick Boat Company sites represented.

VALUE ANALYSIS/VALUE ENGINEERING: VA/VE

The third tool I'd like to cover is Value Analysis and Value Engineering (VA/VE). This tool has been around for almost 80 years. In 1947, after WWII, during the rebuilding of the world and the increased demand for products from the US, many industries suffered from a shortage of labor and materials. Lawrence Miles, a purchasing manager for GE, coined the term Value Analysis to describe a systematic approach to improving products and product performance while reducing costs (Note: The affiliated methods and intent align nicely with Design for Affordability). The initial focus of the effort was to find suitable substitutes for a shortage of product components and materials. This often resulted in improved products at lower cost. Value Analysis is for understanding the performance and cost of a product, while Value Engineering is for improving the product while attempting to lower the cost (Note: Follows the mantra: Faster, Better, more Affordable).

Value Analysis can be applied to both existing products and existing processes. The main goal VA is to understand cost and improve value without compromising quality of performance. Value Engineering is applied to design and development. The goal of VE is to optimize the functions and performance of the product at the lowest acceptable cost by redesigning components, selecting materials and improving processes before production. Although both emerged from a build/manufacture environment, the principles can be applied to services as well. Although both are often used interchangeably, here is a table that compares and contrasts the two when using VA/VE methodology:

Dimension	Value Analysis	Value Engineering
Stage of Application	Post-Production	Pre-Production
	Existing Products	Design and Development
Objective	Cost Reduction	Cost Prevention
	Elimination of Non-Value Add	Performance Improvement
Approach	Systematic	Preventive
	Remedial	Creative
	Function Based	Multi-Disciplinary
Team	Engineers	Engineers
	Cost Analysts	Designers
	Purchasing Professionals	Customer Reps/Marketing
Focus	Identifying Unnecessary Costs	Enhancing Product Value
		Improving Functionality

VA and VE utilize a function-based team approach:

1. Identifying the essential functions or components of a product or process.
2. Evaluating alternatives to achieve these functions.
3. Selecting the most cost-effective and efficient solution.
4. Implementing changes while maintaining or improving quality and performance.

Typically, VA or VE teams meet for one to three weeks, during which they work through the "job plan" as outlined by the standard. According to the Value Methodology standard (Note: see the standard for more detailed information), there are 6 phases to a Value Analysis or Value Engineering:

Stage VA	VE
Information on existing Components	Information Gathering on Potentials
Function Analysis	Function Analysis
Creative (on Alternatives)	Speculation on Potentials
Evaluation of Alternatives	Evaluation of Potentials
Development of Alternatives	Development of Potentials
Presentation on Team Results	Presentation on Team Results

The presentation is the final phase and culmination of any VA/VE event. Just like 3P, the alternatives and potentials are presented to a body of decision-making individuals for determining what to move forward with, and what to decline for implementation. Since changes to product norms will likely be required for instituting alternatives or potentials, individuals such as a product architect, a technology expert, engineering managers and directors, production managers and director, VOC customer rep.s, and other engineers and manufacturing personnel may be involved in viewing and discussing the presentation. Often, the presentation requires more review and more consideration by the decision-makers before moving ahead to implementation.

LET'S SLEEP ON IT – VA/VE

Approximately ten years ago, I was invited in to deliver a VA/VE training program for some of the engineers for the Serta Simmons Bedding Company (Note: Serta was founded in 1931, and Simmons was founded in 1870). With the increase in competition over the past 30 years, and the increase in use of new materials and technologies, Serta Simmons determined that it could use a standardized approach for deciding what new methods, materials and technologies to use with their existing products and what components they needed to design and engineer new products.

Because competitors were offering new functions and features (e.g., foam, automotion, various settings, heating/cooling, etc.), Serta Simmons had to determine what to embrace and what to reject in existing and new products. VA/VE fit nicely into their own requirements and demand to be more competitive. After spending several days transferring knowledge and capability on VOC, Lean Systems and VA/VE, the product engineering teams participating were assigned to projects pre-defined by leadership aimed at improving product performance, reducing cost (product and process) and increasing product quality.

Today, Serta Simmons is still a top choice for bedding. They did not want to offer products with "All-the-Bells-and-Whistles" or offer products that attempted to be "All-Things-To-All-People", but instead, engineer products that meet specific requirements, needs, wishes and wants of specific customer target groups. With their attention to ever improving Product and Process, and their focus on continuous improvement of Quality and Performance, their dedication to Deliver Value that meets and exceeds Customer Requirements (at a price the customer can and is willing to pay), insured they will maintain their position in the industry.

Please keep in mind, the tools featured in this chapter are only but a few available out of the whole set of Affordability Design Tools. For additional information you may want to do some research on: DFX (Design For Excellence), DFSS (Design For Six Sigma), DFM (Design For Manufacturing), DFLM (Design For Lean Manufacturing), DFA (Design For Assembly), DFL (Design For Logistics), DFV (Design For Variability), DFC (Design For Cost), DFI (Design For Inspection), DFS (Design For Safety), DFQ (Design For Quality) … as well as many others.

Design for Affordability applies to Product/Service Design and System/Process Design. This approach ensures that the Value being Delivered (i.e., Product and/or Service) is created, produced and provided via a System of Processes meeting and exceeding customer requirements and fulfilling market demand, in an efficient and cost-effective manner. Appropriate time spent upfront designing and planning, provides a better, faster more Affordable Product, Service, System or Process.

YOUR PATHWAY AND YOUR JOURNEY

At this stage of the journey, your design center is of utmost importance. Whether your value proposition is via a product or a service or both, your UVP (Unique Value Proposition) should be understood and possibly defined at this point. You should have a comprehensive knowledge of all requirements for the target customer base for your UVP. In addition, a price range should be known and customer verified, and the resource expenses required for creation and delivery of Value should be established.

Additionally, the enterprise, systems, processes and procedures for creating and delivering value should be either in draft mode or at some level of recorded documentation (i.e., Value Streams, designs, illustrations, SIPOCs, detailed flows, etc.). The resources required (i.e., means, methods, machines, tools, materials, money and other miscellaneous resources) should be identified and defined. The profile of the people to hire, or those already hired, should be known with planned recruitment, hiring, onboarding and institution practices in place.

Although most often avoided, an HOQ for QFD would be quite beneficial if its in place. Over the years, with 137 clients served, I've only had two full HOQs completed with a third being the one I created while at NCR. In full disclosure, while I emphasize the importance of this tool to all clients, I've only had it fully utilized a few times. It may be because of its complexity and intricacy, but the reluctancy to apply it does dis-service to the full extent of the capability of Affordability.

Also beneficial is the execution of a 3P event to prepare the system for value production and delivery. In October 1989 after NCR spent the better part of a year preparing for Just-In-Time production (without a 3P), it still took another year (i.e., 1990) to adjust and tune the system to be effective. It took the better part of five years (1990–1995) to achieve a level of performance excellence for their J.I.T. or Lean Assembly Plant.

VA/VE should be under consideration for use with new or existing products, services, systems or processes. Keep in mind, VA/VE can be used in conjunction with QFD, VOC, HOQ, 3P and the other methods for the definition and design for Value and the Value Stream.

In closing the Design for Affordability chapter, I can declare, "The more time and effort spent on Design for Affordability upfront, the better the operational performance outcome, the greater the value stream performance, the more effective the costing, and the greater the profitability and prosperity." (Note: Please do not catch that dreaded disease, "Paralysis by analysis and planning". But do keep that healthy practice of continued planning for adjusting and accommodating to new and/or unforeseen circumstances that arise.)

7 Targeting and Aiming

When you see your target, your aim is perfect.

– Woody Flowers

AFFORDABILITY TARGETING AND AIMING

I've noticed that some organizations often shoot at too many targets … and it may evolve to a situation that everything's a target, an in essence, there is no true target. Other organizations have too few targets and too often only one … money. Aiming suffers some of the same dynamics … the most recent "shiny object" in the marketplace attracts attention and becomes the aim. Or, changing aims frequently creates chaos and confusion within the workforce. Across an enterprise, there could literally be millions of variables and conditions to observe, monitor and measure. If you're shooting at all of the millions of possibilities, its common sense to conclude you're wasting your energy and ammunition and likely spending a lot of money that will not return a single dime.

If there is only one target, you're likely missing critical targets, that should be prioritized and addressed. For example, if your primary and only target is based on money and money alone, it's like driving down the road looking in the rear view mirror … you're likely to eventually run off the road (i.e., Money is a lagging indicator of performance. You can't safely drive a vehicle if all you use is the rearview mirror.). It is true, money is important, but Affordability is achieved by targeting and aiming at The Value, Value Delivery, Value Stream Performance, Customer Satisfaction, Customer Requirements, Employee Satisfaction, Supplier Satisfaction, Competitiveness, System Speed, System Quality, Cost, Expense, Spend, Profitability and Prosperity.

The first step is to identify the critical targets necessary to achieve Affordability at the Strategic, Operational and Tactical levels of the organization. System targets should be aligned with Enterprise targets, and in turn, Process and Procedural targets should align with the Systems level targets. People targets should be aligned from "the floor and task" level all the way up to the executive office. Then, all critical targets should be baselined (i.e., measured at the current state condition) to understand the current performance measure, reasonable and achievable targets should be set (Note: If at all possible, with input from the people who will ultimately be responsible and accountable for the results).

The second step is to communicate the targets and aims (i.e., goals and objectives) and institute a system that measures, displays and reviews performance on a regular basis. At the individual and team level, the management level, and the directorate or executive level, a daily 15 minute "huddle" can be established that puts in motion a rhythm addressing current performance, opportunities for improvement, improvement projects and achievement recognition.

DOI: 10.1201/9781003772934-7

The third step is to manage and maintain the performance program, changing targets as necessary and adjusting aims as required. The overarching goal is to increase performance. The focus of performance should be viewed Strategically, Operationally and Tactically. A linkage and alignment should be designed into the performance measurement system.

Strategically, I recommend you begin with the Affordability Scorecard: Customers, Business, Process, People. Operationally, I suggest you start with the Value Stream and Systems that Deliver Value. Tactically, it is prudent to focus on processes, procedures and tasks that align with qualitative metrics (Customers, Employees, Suppliers both internal and external) and quantitative metrics (Time, Quality, Cost). Be sure that there is a connection, linkage and alignment of all targets from the Strategic, Operational and Tactical levels.

The aims established should be reasonable and achievable, and if possible, defined by the people responsible and accountable for the results. S.M.A.R.T (Specific, Measurable, Achievable, Reasonable, Time Bound) goals and objectives should serve to align and aim the people and processes for performance. Successful performance should be acknowledged, recognized and celebrated, while unsuccessful performance should be leveraged to teach, develop, advance, improve and adjust the people and processes for achievement of performance excellence (i.e., Performance Excellence happens when the aims of the targets are met and exceeded. Sometimes referred to as a "bullseye of better.").

The fourth step is to be sure that continual and continuous improvement is in place for targets and the aims of goals and objectives. Continuous improvement is the ongoing discipline that assures reasonable targets and legitimate aims for performance increases and successful outcomes. Targeting and aiming with the proper metrics and measurement systems in place provides leadership with intelligence to adjust course when needed, maintain the course "steady as she goes" or proceed "full steam ahead" when the time is right.

If you know your Value, if you know your Customer, if you know your Cost, you know your targets and your AIM is perfect. At the Enterprise level, your Customer, Business, Process, and People targets are clear (Note: Or, should be clear.). Linking these Strategic targets with Operational, Tactical and Team/People targets, provides for a contiguous understanding of performance. Knowing the baseline measures, understanding the possibilities, setting the goals and objectives, the aim of the current state to get to the future state should be crystal clear.

THE WORST PLACE TO WORK IN THE COMPANY

Almost 30 years ago, I was invited in to help Store Automated Systems, Inc. (AKA: SASI) in Bristol Pennsylvania with issues they were having with their technical service and support desk. SASI was co-owned by four individuals resulting from the merger of a software company and a hardware company to create a computer systems solution company that served the Retail Market Organizations (Their clients spread across the country and included the likes of PEP Boys an automotive company, Rite Aid Pharmacy and several Supermarket Companies). One of the Co-Owners, Jon Heyward, held several titles and assignments as Chief Scientist, System Designer,

Software Architect and Technical Guru. He asked me to work with him to address performance issues of the SASI Technical Service and Support Desk (AKA: The Help Desk).

After spending a good deal of time studying, understanding and assessing the current state of the Help Desk, I could summarize the conditions and situations in a phrase often quoted around the company: "The Help Desk is the worst place to work in the company." Here are a few quotes I heard from people on the team: "there is no way to get transferred or promoted out," "there's no way to leave unless you quit or die," "we don't get any respect nor help from anyone else in the company," "this is the worst place to work in SASI."

I easily determined that the amount of time necessary to turn the group around was way too much for Jon to spend and still be successful with all his other assignments. With the merger, Jon came from the software side of the company, as did the CEO, and with their strong support I was able to make great strides from the project's outset. The strategy and direction was based on improvement, the leadership was onboard from the beginning, the people were more than ready for change, I had tools and resources ready to go and the design and plan had to be created.

The team was staffed with 18 associates and operated for 24 hours a day (i.e., three shifts: 8:00AM–4:00PM, 4:00PM–12:00AM, 12:00AM–8:00PM). They were physically arranged in what I call a "cube farm," or a cluster of typical office cubes. They were situated near the cafeteria area away from the rest of the employees because, as I was told, their "discussions over the phone disturbed others." First shift and second shift had a manager, a team lead and six associates manning the phones. Third shift had two associates that served as "clean up" from the calls of the day. Most of the calls started coming in at 9:00AM and ended around 10:00PM (Note: Their customers were scattered across the US from the east coast to the west coast. Most of their east coast customers opened at 9:00AM EST and their west coast customers closed around 12:00PM EST, 9:00PM PCT.). As I observed the operation, their heavy call rate seemed to occur 9:00AM–12:00PM and 9:00PM–12:00AM (i.e., The opening times and the closing times for their customers, from the east coast to the west coast). During peak calls times, the area was a bit chaotic. They only had two technical manuals (in notebook physical form) to work with, and when an associate needed a manual, they'd make a request, and the associate with a manual would literally pass the manual "over the cube wall." Their office area was quite sterile, except for the personal pictures and mementos inside each associates cube.

One of my first activities involved interviewing the associates in order to understand their view and perspective of the work and their job's purpose. I sought to determine and define their personal view of the mission and aim of the Help Desk. Inherently, they knew their role was to provide customer technical support. However, I quickly realized after asking each associate a few questions, their current state perception bordered on "clueless." Here are some of the questions I asked in my one-on-one interviews with them:

- About how many calls do you answer each day? ("Don't Know")
- About how long does each call last? ("Minutes to Hours," "Don't Know")

- On average, how long does it usually take to get a solution to the problem back to the customer? ("Sometimes minutes, sometimes hours, sometimes days.")
- What type of support do you receive? ("The manual, the lead, the manager, that's all.")
- How satisfied are you with the job? ("Not Much," "Poor," "Discouraged," "Frustrated")
- How do you know at the end of the shift, you had a 'good day'? ("I didn't get yelled at")
- What kind of personal growth do you expect? ("This is a dead-end job." "This is the worst place to work in the company.")

I also interviewed the managers and team leads, and asked the same questions with a few additional queries. The managers and team leads were much more technically astute than the associates and a bit more upbeat about their jobs. They suffered from frequent interruptions because they had to constantly field customer questions that were beyond associates' competence (Quote—"I'm so busy fielding calls, I don't have the time to train anyone."). Surprisingly, they were passionate and dedicated to customer service but suffered from a poor design.

I asked the managers and team leads one specific question: "At the end of the shift, or the 'end of the day', how does the team know if they're winning or losing?" The typical answer, "They don't." I also asked: "How many calls does the Help Desk take in a year, in a month, in a day?" The consensus, "We don't know." Finally, I asked, "On average, how long does it take, from the time a customer calls, to the moment they receive their answer?" Unanimity: "We don't know."

As a result of the interviews, and the additional observations, I was able to "paint a picture" of their current state. Picture This: You live in a sterile cube farm at work. You don't know what is expected of you each day except to try and keep the customers satisfied. You haven't been trained. You don't have your own documentation nor any equipment readily available to research, discover and test customer solutions. You receive little support outside of your lead and the managers. There are no goals, no stated purpose and mission, no direction, little communication and you're mostly avoided by the rest of the company. In fact, you don't what's expected of you and the rest of the team regarding, number of calls (each day, each month each year), length of calls, length of time-to-response, quality of response (i.e., Did the response solve the problem?, Was a call back required?) and there were no metrics and measures for the team to target and aim. At the end of the day, 'you didn't know if you were winning or losing … except when you didn't get yelled at that day'.

What I really liked about the group was that they had a sense of urgency to change and transform. They were actually surprised that they received leadership's attention, and they would get a chance to improve their workplace. The initiative was a long-time-coming response to the managers' and team leads' requests for help. The 'big boss' Jon was technically brilliant with too many jobs to perform, and no time available for a group transformation. This project offloaded his biggest headache from it being a 'pain' to being a situation he could support and participate in. It was a pending win-win all the way down the line from the top executive to the everyday associate.

Knowing that there were no baseline metrics and measures, I had the opportunity to help them establish reasonable and achievable targets. I was able to assist the team define their performance targets and goals. Their consolidated identification of targets initially came to: Call Length, Time-to-Resolution, Call Quality, Call Response Quality, Daily Count of Calls (Per Associate and Per Team), Call Cost with written notes on what additionally was needed, what was missing, what was required and what support was lacking in order to correctly respond to a customer call. They now had targets to aim at that proved their success and performance, so we set up a highly visible performance communication board to display their performance (Note: This was the first performance board on display in the company. It attracted a lot of attention and people were coming in to see how things were going with the Help Desk). At first, the group was reticent and reluctant to "show their work." After they began to experience quick wins, and attract positive attention, their negative perception became a positive reality and sense of pride.

As they posted call notes that identified their needs (e.g., improved work environment proper documentation, equipment for test and evaluation of solutions, engineering support for difficult problems, product training, etc.), I was able to pursue additional quick wins to create more momentum and leverage new found success (Note: It is true that "Success Breeds Success"). In parallel, with support and manager participation, we were able to institute Lean practices via Kaizens, Workplace Improvement and Systems development (i.e., Lean Thinking, Value, Value Stream, Process definition, development, management, etc.).

At the end of the first year, during the leadership review of the results, the leadership team wanted more development in the Help Desk area since they began to realize, that group was a direct point of contact with customers, that directly affected Customer Satisfaction. The second year of activity involved instituting more advanced targets requiring new metrics and measures (i.e., Customer Satisfaction, Employee Satisfaction, "Supplier Satisfaction" being leadership, engineering, marketing and sales.). As the team advanced, the aim for the targets changed to reflect new found performance (Note: No aim was set that made it impossible to achieve given the current state at the time.).

By the third year, the team was performing as good as any Help Desk I've ever experienced. In less than three years, they went from a hopeless failure condition to the best performing 'department' in the company. Their advancement in knowledge, experience and success provided an environment that permitted them to design their own team workspace. The open approach increased communication, improved problem solving and accelerated their success. Originally, each individual was in a do-or-die situation, and within a few years, the individuals operated as a high performing team.

As you can see from the "before and after" diagram below (Figure 7.1), their physical appearance and the basic performance dramatically changed:

As you can see, as the initial number of calls increased, The Help Desk Team was were able to reduce their average response time with fewer people, improve the quality of their responses and lower their costs (Note: Since their ultimate design required only one manager, the other manager was promoted to be a software engineer as he was "promised" when he was hired. One team leader was promoted to

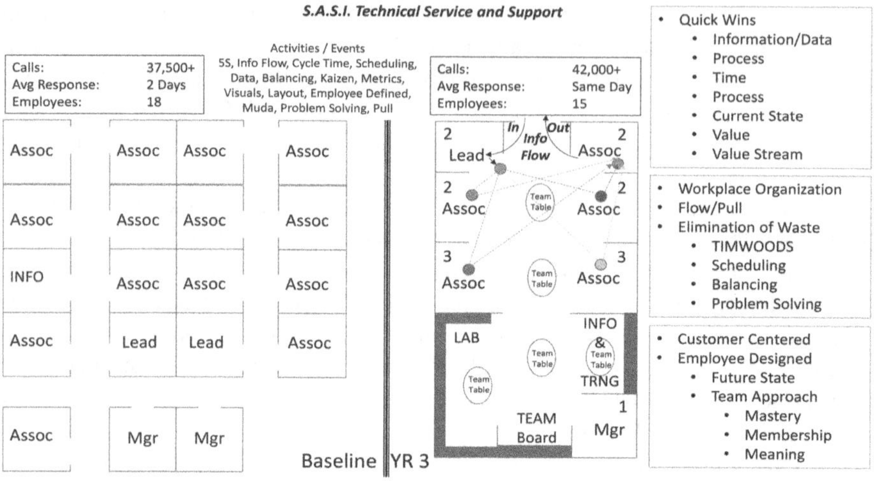

FIGURE 7.1 A three-year transformation from "Cube farm" to dynamic integrated solution system.

serve in marketing in concert with her degree and a "promise" when she was hired. One associate was promoted to team leader per recommendation of the manager. Another associate was promoted to be a manufacturing lead, and again, it was a job he was "promised" when he was hired. Not surprising, the Employee Satisfaction score went way up!

Its also worth noting, the design of the Help Desk, changed from a "cube farm," to an integrated team environment with a new culture and a proud-of-my-work attitude. They instituted training, increased documentation and training (Note: with a documentation library and a training area), and created a lab with current the version of equipment and software for the purpose of testing and validating proposed customer solutions.). What's not evident, is the access the Help Desk had to hardware and software engineers for addressing complex customer problems.

And to think of it, this all began with the establishment of targets and aims "defined" by the associates (Note: The "boss," the managers and the team leads were all involved in the definitions, communication, and improvement throughout the three project years.). It is often asked, "How long does it take to transform a group?" In this case, it took three years, beginning with targeting and aiming as the first activity and event.

Target the critical factors for your success, your Aim should be directed toward goals and objectives that are achievable. As you perform measure the outcomes and adjust accordingly. Adjust and improve your aim consequently as the outcomes, accomplishments, achievements and results are realized. Add targets as necessary and manage your measurement system to avoid too many or too few targets and eliminate targets when they are no longer important or critical to performance.

Target Affordability with the aim of addressing the factors and executing the formula of success. Target the performance of the critical elements of Affordability: the

Value Stream, Value Delivery, the Cost of Value, Customer Cost and other Customer Dimensions, Product and/or Service, Business Performance, Process Performance, People Performance and Growth. Apply additional targets as necessary to obtain a full perspective of performance.

YOUR PATHWAY AND YOUR JOURNEY

If you're just starting a business, it's a common reality "Cash Flow is King." Major start-up metrics involve money measured as income/revenue, cost/expense/spend and profitability/prosperity. For the Value you provide in terms of products and/or services, I suggest you measure fit with Customer requirements (i.e., requirements you meet, requirements you don't meet). It would also be prudent to include process performance in terms of time, quality and cost, and people performance of Customer, Employees and Suppliers.

If you've been in business for some time, your performance scorecard should be improved to include the success metrics and measures I've mentioned in this chapter. Consider eliminating metrics and measures that don't fit the Value you provide and the Value Stream you use to Deliver Customer Value.

Always keep in mind, if you don't keep score, and you don't maintain a scoreboard for communicating performance, you'll never be able to sustain your business, nor grow and increase your market … you'll never know if you're winning of losing. But first, be sure you define your targets from the Enterprise level and link the metrics and measures throughout the operational and tactical levels to the team and people performance dimensions. For Affordability qualitative targets should be known is the areas of Customers, People, Suppliers and Partners (Note: The aim is on Satisfaction and Delight.). Affordability quantitative targets should be based on time, quality and cost. Variants in each category should apply directly to Value, Value Stream and Value Delivery. All other targets and aims should be linked and connected with all of the primary targets. Hence, the "aim is perfect."

8 Planning and Implementation

Plans are useless, planning is indispensable.

– Dwight D. Eisenhower

Ideas are useless unless used. The proof of their value is in their implementation. Until then, they are in limbo.

– Theodore Levitt

AFFORDABILITY PLANNING AND IMPLEMENTATION

The duration, difficulty and dynamics of planning and implementation vary depending on the type of activity, the content involved, the complexity of the subject matter and the individuals participating. Planning and implementation for Affordability may involve a product, a service or a process, but it may also entail projects, training workshops, strategy deployment, mergers and acquisitions or even verification and validation of a concept or idea. Often, there's not enough planning for critical implementations, or at times, there's too much planning for simple, just-do-it achievements. One dangerous scenario I've encountered sometimes happens when you gather a team of highly capable and intelligent professionals in a room, present a problem that requires a solution, and without any planning, they collectively "jump-to-conclusions," quickly come up with a solution and begin to implement prematurely, quite often ending in failure. Success often depends on balance and time when it comes to that amount of planning as compared to the quantity of time.

The balance of planning and implementation is not always a true balance. For example, when I prepare learning sessions or workshops for knowledge transfer and capability development, its usually a 2-to-1 ratio of planning to implementation (i.e., two days of planning for every one day of delivery). When I create a simulation for hands-on, experiential training, it has taken one week of planning and creation to deliver a half-day simulation activity. One of my first projects for AT&T took the better portion of a few months to plan for a ten-year implementation rollout of a strategy to return the company back to the communication business. Initial planning is done and readiness for implementation occurs when you or your planning team is satisfied that all considerations, variables and possibilities are address and satisfied.

The timeframe for planning is not finite. Once a plan is created and the implementation has begun, it is most common that the planning will continue throughout implementation if the planning assumptions, conditions and implementation variable shift and change. When an idea or concept is somewhat realized, without a plan for implementation, and ongoing planning during application, such discoveries,

DOI: 10.1201/9781003772934-8

creations and inventions fail to be realized. I've heard people say, "Planning for implementation is both an art and a science." I have discovered that successful implementations contain distinct elements that go beyond balance, time and technique for success to be achieved.

I often look for several elements to be in place for me to have the greatest possibility and probability for success. Initially, there must be a good reason and intention to take on any action or activity. To begin with, the establishment of purpose, aim and goal, accompanied by vision, mission and direction should be in place before commencing. The identification and presence of leadership must be established. The right people, resources and tools should be available as required. And finally, the solution design with the proper implementation plan should be known. I refer to this as the Success Formula:

Success Formula = Purpose + Leadership + People + Resources and Tools + Design and Plan

The purpose is the reason for planning and implementing. It is the reason for putting an effort forth. It could have emerged from a sense of urgency, a requirement from a customer or market segment or as a result of a vision, mission and direction set by a leader. Without purpose, activities can be judged as "busy work" with no value. Purpose can serve to set direction, align resources, motivate people, communicate the message and aid in execution of the plan.

Leadership and a leader must assume responsibility and accountability that the plan is accomplished and the implementation is achieved. Without leadership there is often chaos and confusion. With strong leadership final execution will be realized. Participatory leadership ensures that the direction and alignment requirements are trusted since the leader is directly involved (i.e., Behaviors: Model the Way, Inspire a Shared Vision, Challenge the Process, Enable Other to Act, Encourage the Heart).

The right people are necessary to complete the implementation. People employed to carry out the plan should have been included in the creation of the plan. A team of motivated skilled talent, intimately cognizant with the plan, should be the "execution squad." The implementation team should be comprised of competent individuals that believe in the solution to be implemented. Resistors and 'nay sayers' should not be included in the planning nor implementation efforts.

Effective resources should be made available for the efforts and activities of the people engaged. Resources include materials, machinery, tools, methods, manpower, money and any other ingredient that may aid in planning and implementation. Lack of resources, or missing necessary resources destroy any team's efforts. Leadership is responsible to see to it that all resources necessary are made available.

Tools for planning and implementation are readily available for addressing products, services, processes, projects, strategic deployment, workflow, personnel recruiting, hiring and onboarding, etc. Dr. Ishikawa, a famed quality guru, identified seven tools for gaining understanding of a problem before even planning for implementation of a problem's solution (i.e., seven tools: Matrix Diagram, Affinity Diagram, Interrelationship Diagram, Prioritization Matrix, Tree Diagram, Process

Decision Program Chart, Activity Network Diagram … not to mention the Ishikawa Diagram of Fishbone Chart).

A detailed solution design should be delivered from planning for implementation. The solution implementation design should contain all facets and aspects that are required to meet the purpose and of the organization. The design should be documented and detailed in such a way that every component is clearly understood by every participant.

Above all, the plan for implementation should contain: The Why, The What, The Who, The When, The Where, The How. The final implementation should contain all the components of the plan. Planning should continue throughout the implementation to insure that all factors are included and all necessary adjustments are made throughout the duration of implementation time.

Even with all the elements of the "Success Formula" in place, there is always that chance of failure. A Victory should be celebrated, but a "Loss" should be a "Learning." Regardless of the outcome, the practice of capturing and studying "Lessons Learned" is often an ignored step after completion of an event or activity. Try to instill this discipline into every action you take and every activity in which you're engaged. 'Re-Living the Journey' (From "Incredibly American" by Lew Hatala and Marilyn Zuckerman) for the team to re-call and discuss success and opportunities for improvement, and 'Sharpening the Saw' (From "7 Habits of Highly Effective People" by Stephen R Covey) for learning, growing and improving, are two often forgotten techniques for developing and growing teams.

Planning and implementation of Affordability 2.0 requires that the foundation and framework of Affordability be used for products, services, systems, processes and initiatives. For each and every Affordability Solution the following critical Affordability Factors should be considered:

- The Goal(s): Linked Strategic Goals, Operational Goals and Tactical Goals should be established for Value (products and services), for the Value Stream, for Customer Satisfaction and for the improvement of Business, Process and People objectives. Goals for individuals and teams should be related to tactical, operational and strategic objectives. Measurement systems should be in place for displaying, monitoring, reviewing and discussing current performance.
- Leadership: Affordability leadership has 10 dimensions: setting direction, aligning resources, motivating people, communicating the message, executing the plan, modeling the way, inspiring a shared vision, challenging the process, enabling others to act, encouraging the heart.
- Strategy: As related to goals and leadership, strategy determines the direction and aligns resources to achieve objectives. Strategy can be set for the organization, the Value Delivered, the market, etc. A plan to execute to achieve strategy should be declared and deployed (i.e., Strategic Plan).
- Enterprise and Systems: Systems of the enterprise are the operational means and methods of executing the purpose of the enterprise. Systems are comprised of the processes, procedures and tasks that accomplish the work and operate the Value Stream.

- Design for Affordability: Using the features and functions of Affordability as a guide, Design for Affordability applies to the Value (products and services), the Value Stream, the People, the Processes and the Performance of the organization.
- Targeting and Aiming: Once Goals and Objectives are established, with specific targets defined and aiming to hit those targets can take place. Specific targets at the strategic, operational, tactical and individual/team layers can be established, measured, monitored and used as the aim.
- Planning and Implementation: ongoing planning and implementation applies to strategy, products and/or services, systems and processes, people, problem solving, improvement and any activity that requires the use of resources.
- People: The workforce, the drivers of process, the individuals who create, produce, deliver Value and those who support all of the Value Stream.
- Process: The means and methods by which the enterprise operates. Integrated processes make up systems, and systems makes the enterprise that Delivers Value to the Customer.
- Performance: The measure of how well or how poorly actions, tasks and functions are executing designed activities using specific metrics to monitor outcomes. Performance can be used to make adjustments, correct inappropriate action or direct action and activities.
- Improvement: In essence, "Make Better." Improvement is an Affordability factor that applies to the Value, the Value Stream, the Customer, the Cost, as well as all of the other factors.
- Sustainment: Ongoing support and maintenance directed toward preserving longevity and purpose.

Affordability Solutions may not formerly contain every one of the ten factors listed above, however, all solutions should use the factors for framing considerations of solution aspects and decision making. Affordability Thinking utilizes the Affordability Factors as a checklist and potential outline for the formation and creation of solution realization. I have two examples I'd like to share that specifically pertains to the use of Affordability Factors for planning, preparation and implementation (i.e., A Hybrid Scheduling Approach for NASA and A Lean Initiative Kick-Off for Virginia Blood Services—VBS).

A HYBRID SCHEDULE ... AGILE WATERFALL FOR A LAUNCH TO THE MOON

In August 2023, I was invited in to assist the Artemis II NASA Program to work with them on their approach to scheduling. The Artemis II Program was created to fly a crew of astronauts around the moon and back to earth. Originally, back in 2011, the launch date was set for some time between 2019 and 2021, and afterwards the launch date was delayed to 2023. It was realized in the summer of 2023, that the launch would not happen during that year. Traditionally, from the beginning of the NASA program back in 1958, the scheduling approach used was the traditional linear waterfall model. More recently, an adaptive iterative approach was emerging within the ranks, especially for the software development groups. By the end of the summer of 2023, the leadership of the program realized a contention and conflict between the various groups when it came to scheduling. Most of the organization, the primary functions that emerged out of the Apollo and Space Shuttle programs were using the long-standing traditional approach which is linear and sequential for scheduling missions involving literally millions of distinct actions and activities. The rest of the mission team was using an Agile approach which tends to be more flexible, collaborative and iterative. With two competing scheduling methodologies it was clear to the top leadership that some form of integration was necessary.

I was alerted at the end of August about their requirement, in which they requested that I prepare and deliver a program for the leadership, a group of functional managers and a select collection of professionals on September 26–28 and December 6–7 focused on achieving a better approach to scheduling. I found it a bit amusing, that historically, NASA spends years scheduling for a program, and they gave me a month to deliver a program on scheduling (Note: Granted, their missions were, and are, much more complex and complicated than the one assigned to me.). The difficult challenge was how to address both sides and bring them together on a consensus of how optimize the scheduling methodology.

The 1-Day Leadership session (held on September 26) was a distillation of the 2-Day session planned for the managers and professionals (held on September 27–28 and December 6–7). Since the all of the team was very familiar with the customary waterfall method, with a subset of the team using flexible, collaborative, iterative alternative, I incorporated a substantial amount of Lean and Agile content in the beginning in order to bring all parties up to a common understanding of both "competing methodologies" by the end of the workshop. Those using Agile scheduling were familiar with 'Waterfall', but most of those using 'Waterfall' were not acquainted with Agile. As you might expect, the "Waterfall People" favored 'Waterfall', while the "Agile People" preferred 'Agile'. There was a great amount of discussion (and argument) over which method was best. However, over time, both sides progressed toward the end goal. By the end of the workshop, the two were blended and the 'Hybrid Methodology' prevailed. The top leadership fully supported the Hybrid Methodology and were more than ready to discard the conflict and confusion caused by the two competing methodologies.

During the sessions, some tangential discussions took place regarding launch readiness. Although in 2023 the administration was pushing the ASAP attitude with a sense of urgency to launch and take credit, the young engineers, as well as the seasoned veterans, all verbalized, "We're likely two and a half years away." Now, as we look back, the launch preparation and execution team knew in their hearts the reality that they faced. Once again, the adage was proven, "When the team truly believes it, it will be done." Even with unforeseen, unexpected and unplanned delays, the hybrid planning methodology launched Artemis II on April 1, 2026 from NASA's Kennedy Space Center, two and a half years later.

THREE-DAY WORKSHOP: PLANNING AND IMPLEMENTATION RESULTING IN A $37,000.00 R.O.I.

About 15 years ago, I was invited in to Virginia Blood Services (i.e., VBS) to serve up a Lean Workshop that provided them with the first step of their Lean Journey. The VBS Value Stream involved delivering whole blood, blood components and blood plasma to Southern Virginia Healthcare Establishments that existed below a line drawn west to east through Fredericksburg, from West Virginia to the Potomac River including the areas of Charlottesville, Roanoke, Blacksburg, Richmond, Williamsburg, Norfolk, Newport News, Virginia Beach. etc. Northern Virginia (Often called N.O.V.A.) was supplied by the American Red Cross (Their "Competitor").

The initiative was concocted by one of their Senior VP's in an effort to begin to adopt Lean practices throughout VBS. The SVP, her administrative assistant and I spent a three-week span of time planning for the three-day workshop. From my Japanese coaches, mentors and partners, I learned that if you spend a lot of quality time upfront in detailed planning, the implementation will turn out better and less time will be wasted during the execution of the plan. During the first week we concentrated "The Why" (i.e., Why we were there, the purpose, the intent, elements and the expected results). Week two covered "The What" (i.e., What we were going to do, the topics to be covered, the exercises for the group, the expected results and outcome.). The final week before implementation, we address "The When and Where" (i.e., When we would perform specific activities and events, and Where our actions would take place.). "The Who," or the participants were pre-selected by the SVP (i.e., The Participants were 15 employees who were forward thinking and passionate about improving the workplace.).

The VBS Value Stream consisted of primary processes: Setting Up for a Blood Drive, Executing a Blood Drive, Processing the Blood collected at a Blood Drive. The group was comprised of five professional from each process. The first day was dedicated to knowledge transfer and training regarding Lean Practices, and the development of each team's current state process map. Day 2 consisted of the remaining Lean Learning (including identifying waste), the actual identification of waste observing within each team's current state process map and creating a rendering each team's future state map (i.e., without the waste) and with a plan to execute a mini-Kaizen event on Day 3 for waste elimination. The last day started with setup for each of the three planned Kaizens and execution of a mini-Kaizen event for each team by the end of lunchtime (i.e., The Three Kaizens were performed by the three team of five individuals: The Blood Collection Set-Up Team, Blood Drive Blood Collection Team, The Blood Lab Team). Each Team had to prepare a presentation of their Kazen results and present their accomplishments to the top executive leaders of VBS.

At the end of Day 3, the top executives of VBS were brought in to view, observe and comment on the outcomes of the three Kaizen Events. Each team performed in the order of the Value Stream: Blood Drive Set-Up → Blood Drive Blood Collection → Blood Lab. The presentations were excellent and full of accomplishment. The senior leaders were quite impressed by the amount of work done in just a short period of time. Accomplishments included: 15 people trained on Lean Thinking ready for implementation, measured process improvement in all three areas (i.e., increased capacity, increased speed, improved quality and decreased cost), the reduced cost amount to more than $37,000.00 collectively over a period of the ½ day Kaizens (Note: In the beginning, I promised a cost reduction estimate of $12,000.00 to $15,000.00, that would serve as an annual annuity of cost elimination going forward. The cost reduction alone paid for the Workshop and yielded a substantial R.O.I.), and the final deliverable was a list of improvement opportunities to serve as targets for Lean Improvement Events throughout that year.

Planning and Implementation for Affordability targets Faster, Better, more Affordable systems and operations, resulting in greater performance, profitability and prosperity. Affordability combines best practices of the past, with proven tools and techniques, in a unique and creative way of thinking within the Affordability foundation and framework.

YOUR PATHWAY AND YOUR JOURNEY

Planning and Implementation should be established even before any iota of Affordability is realized. Always start with a full, complete, in-depth of the subject matter before attempting to accomplish anything. "Jumping In" or "Jumping to Conclusions" most often ends poorly, especially in business. Planning for a product or service, planning for a customer or market, planning for a change or transformation are all mandatory before implementation.

This very point in time should be your moment of planning to plan, unless you have a plan or are in the midst of preparing a plan. After this point in your evolution, planning before implementation should become a regular behavior of your leadership style and behavior (Note: Include continued planning during implementation to consider and embrace mandatory plan changes.). Some organizations have templates and standards for planning and implementation. Examples of such templates include: Standard Problem-Solving Process, Strategic Planning Process, Product or Service Planning Process, Project Management Process, Invention and Creation Process. As organizations grow, standard approaches save time, resources and money (Warning: Some fluidity and flow are always necessary, too much strict rigidity and inelasticity tend to have a negative effect. The right balance needs to be established. Bruce Lee's Philosophy states it best, "Be Water" (Disciplined, resilient, trained, practiced, flexible, adaptable.).

Some level of planning should always proceed the action of implementation, even for redundant or daily activities. For example, before my daily beach bike rides for exercise every day, I consider temperature, wind direction, wind speed, tide, storm considerations, sand conditions, distance, route, etc., before riding at a specific time, using a specific route, and deciding a specific distance. Once planned, I ride (Note: Each day varies depending on the current state of the variables of the implementation.).

As a leader, your planning or your team's planning, should occur regularly, with consideration of adjustment when critical conditions change. Strategic Plans, Operational Plans, Tactical Plans, Team/People Plans change at a frequency as dictated by the situation at hand and the impact of variables that may change. Typically, the "higher level plans" change less frequently, while Team/People Plans may change daily, or even hourly. Build planning into your culture. Practice planning daily. Change plans when necessary, as dictated by current critical conditions. The performance of the implementation plans is determined by the current status of the targets on the scoreboard or scorecard from strategic, to operational, to tactical to team/individuals. Correct when necessary and adjust the plan accordingly.

If you're planning for the implementation of Affordability, Value and purpose proceeds everything. Value is determined via Customer requirements and expected price. Strategy, systems and structure can be derived by understanding the Value Stream for the Delivery of Value. Processes can be defined and manpower required to drive those processes can be determined. Materials, machines, means and methods can subsequently be designed to support the Value Stream. With the fundamental elements complete the plan can be finalized and, after validating and verifying the contents of the plan, the implementation can commence.

If you're strategically planning for an organization or a product or a service, the purpose, Value, Customer, Cost and AIM should be well understood and defined. You may opt to use a template and flow mentioned in an earlier chapter, or you may select a form or format that applies to your intent. In either case, a standard approach is highly recommended. Implementation of a strategic plan usually encounters some unexpected roadblocks and challenges. As it is with any implementation, accommodate the new requirements, make adjustments, plan for the path forward and continue. Planning and implementation are ongoing until the implementation is complete.

9 People

To add value to others, one must first value others.

– John Maxwell

Try not to become a person of success but rather try to become a person of value.

– Albert Einstein

ALIGN PEOPLE WITH THE PURPOSE TO DIRECT AND MOTIVATE THEM TOWARD VISION, MISSION AND GOAL

Often, when I've asked the question, 'What is the purpose of the organization?'. Many times, the range of responses include: "*To make money,*" "*To protect the company,*" "*To Win,*" "*To beat the Competition,*" "*To make the boss happy,*" and, before Affordability, I never heard the answer, "*To Deliver Value.*" The aim and the purpose of Affordability is to Deliver Value to the Customer at an acceptable price. If Affordability is the framework, the alignment and motivation of the people should be directed toward that main goal, purpose, mission, vision and objectives. On June 25, 2024, Forbes published the article: "The Power of Purpose: Aligning Teams for High Performance," (by Jill Bornstein), that can be summed up in four lines:

> Many organizations have a company mission, vision or purpose, or some kind of combination of the three. A team has to define explicitly for themselves where they fit into this big equation. What is it trying to achieve and for whom, and what does a successful outcome look like? Everyone needs to be on the same page about why the team exists and what its purpose is.

The purpose is the reason for being. The purpose is the driving force for the work of the organization. The purpose is the overarching guiding objective for the people to accomplish. To achieve the purpose, the people must be organized using an organization design and architecture that can carry out the necessary work.

The direction for the architecture and construction of an Affordability organization is simple: Staff to the Value Stream.

Suppliers →	**Value Stream**	**→ Customer**
	From Value Creation to Delivery	
[	Value Stream Support Systems	]
[	Business Support Systems	]
[	Leadership Support	]

DOI: 10.1201/9781003772934-9

Within Affordability, there are four distinct roles to play. Those who are directly involved in the Value Stream from Value Creation to Value Delivery. Those who provide Value Stream Support throughout the entire Enterprise. Those who provide Leadership Support for the people of the organization. Business Support responsibilities range all the way from Suppliers and the Supply Chain to Customer Delivery and Customer Sustainment. Business Support Systems (e.g., Finance, Human Resources, Legal, Technology, Facility Maintenance, Strategic Planning), focus on supporting the operation, the execution of the enterprise systems for the success of the Value Stream and all Value Stream personnel. Leadership support should be applied to the management of all operational and tactical processes and procedures and also applied to leading the people of all the work teams and individuals of the institution.

Staffing for Affordability aligns the people with their purpose in the organization. It's not a departmental design, nor a matrix construct, nor a cross functional structure. Staff to Affordability is a new paradigm that applies people to their role and responsibility as it relates to the Value Stream. It is key to clearly define each role with its responsibilities throughout the four main architectural facets of Affordability. Hiring and onboarding is based on the individual's capability to fulfill the role and fit within the aspects and attributes of Affordability. Employing for Affordability is a new paradigm (Figure 9.1).

The Affordability Model applies to all of the dynamics within the Enterprise:

Using the model, the resources for the "Value Zone" are the people responsible for creating, producing and delivering Value to the Customer supported by those people who serve Value operations. Support functions in the "Value Zone" provide whatever sustenance is necessary to maintain the performance excellence of the Value Stream. Also included are the people who support suppliers and the supply chain.

The "Customer Zone" requires resources to support customer care, customer technical support, customer training, customer planning, customer marketing and sales. Any and all customer functions would be positioned in this zone and responsible for all of customer support and anything that has to do with the customer.

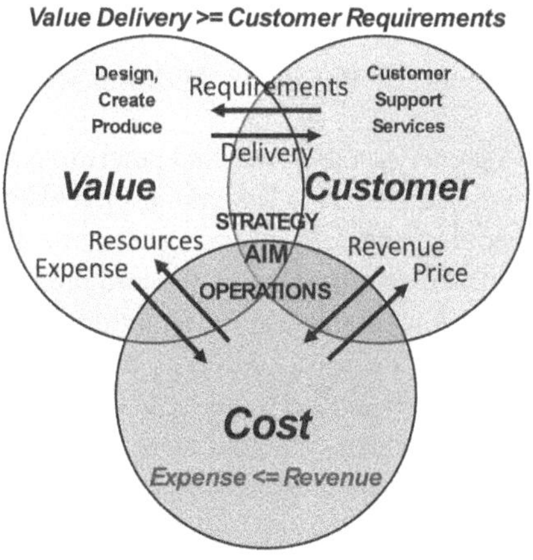

FIGURE 9.1 The dynamics of affordability.

The "Cost Zone" would, of course include financial functions, as well as the sourcing and development of resources ('H.R.') and the technical support that provide technical knowledge and expertise. Other support resources (e.g., legal, maintenance, installation, security, facilities, training, logistics, etc.) can be positioned to support the Value, Value Stream and the "Customer "Zone."

The "AIM Zone" with leadership, management and administration supports the entire realm of the Affordability Enterprise. From strategy and strategic planning, to communications, to operations and other 'corporate functions' the AIM keeps the alignment, providing the direction, motivation, decisions and execution of the plans, policies, tactics and corporate stratagem.

Leadership's overall goal is to energize the people to perform at a level of excellence aligned with purpose for the Delivery of Customer Value that meets and exceeds requirements to satisfy, delight and elate the Customer. Leadership's role is to manage the things and lead the people. For leadership to embrace their role and accomplish their responsibilities, leaders must be developed and have the means to perform. Affordability leaders are best developed by Affordability Mentors and Coaches (i.e., Other Affordability Leaders), prescribed leadership training and experiential learning activities aimed at grooming and preparing individuals to lead teams and people in support of the Affordability Enterprise from Suppliers, the Value Stream, Value Delivery, the Customer to Value Stream Support and support of the extended organizational operational administration functions. At the core, leadership must adopt the Affordability Theory that people drive the systems and processes that result in performance and profitability. This is best illustrated by the 5 P Model, sometimes called, "The Affordability Engine."

PEOPLE AND THE 5 P'S (THE 'AFFORDABILITY ENGINE')

The core of the Affordability Engine is Purpose (i.e., Deliver Value). 5P is an enterprise system where people drive the processes that result in performance which leads to profitability and prosperity that cycles back to the people. It's a powerful cyclical engine (Figure 9.2).

5P's (**P**urpose: **P**eople → **P**rocess → **P**erformance → **P**rofitability)

People are the fuel, the energy, the power and determine the efficiency and effectiveness of the Affordability Engine. It's leadership's job to create a culture and environment that permits people to perform at their best possible performance echelon. It's also leadership's job to be sure that the people are prepared, knowledgeable and capable with the processes, procedures, tools, techniques, means, methods, materials and support for achieving success. Direction, alignment, motivation and communication must be continuously tended to for assuring productive execution of their jobs, tasks, assignments and projects. A qualitative metric of employee satisfaction, and perhaps delight, must be measured, monitored, communicated, groomed and celebrated.

People are the primary input influencing the efficiency, effectiveness, speed, quality and cost of the process. Processes often require machines, tools, materials,

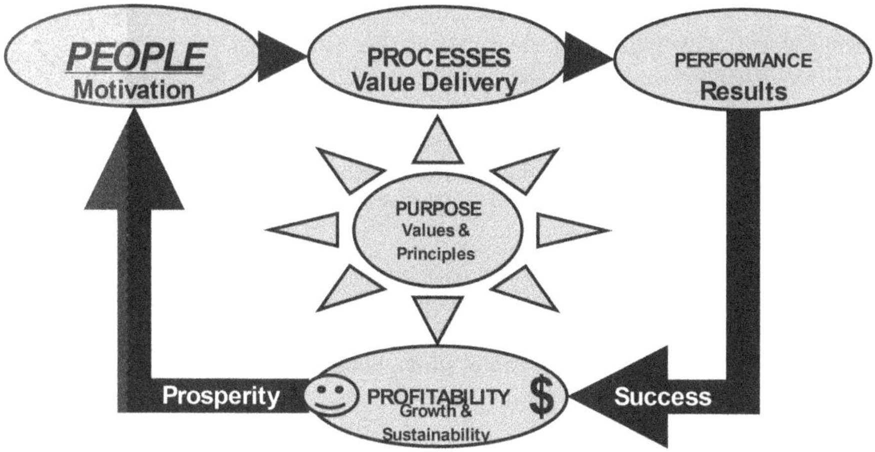

FIGURE 9.2 The 5 P's.

methods, money and management involving people in one way or another. Process problems should be solved by the people. Process outputs and results are affected by the people. All systems and processes should be aligned with the creation, production and delivery of Value.

People's input to the Process(es) result in output, and when measured is Performance. Performance can be measured in qualitative terms (Note: Customers, People, Suppliers, Partners) and quantitative terms (Note: Quantity, Quality, Time, Cost, etc.) to determine the degree of success at the system and process level. At the enterprise level, performance can be scored in the four categories of Customer, Business, Process, People. At the tactical, task, procedural and people level metrics can range according to the measurement that best describes success for each element.

Performance leads to Success and Profitability. Of course, poor performance negatively affects Profitability as much a great performance precedes excellent Profitability. The success and profitability of an enterprise is directly related to the People → Process → Performance linkage. Profitability yields prosperity, and prosperity is the economic outcome that directly influences the Enterprise and the People.

As this engine operates in a cyclical perpetual manner, it goes fast, gets stronger and grows as prosperity increases, but also slows and fails if its driver is failing which impacts the overarching processes related to value delivery, thereby impacting performance and success. As profitability falls and prosperity declines the engine seizes up and stops. However, motivated people drive high performance processes leading to performance excellence and success increasing profitability and prosperity, achieving greater and grander results, consequentially achieving growth and greater wealth and fortune.

Purpose is the core of this cyclical engine. Purpose is the reason for existence; the value being provided and the worth of the product and/or services of the Enterprise. Purpose also includes principles, norms, core values, vision and mission that clearly identify the identity, culture and strategic direction of the organization. As the 5P engine becomes more powerful, the culture gets stronger, and robust norms of achievement, accomplishment and excellence are established.

Although Purpose is a critical point of influence in the 5P Model, the People component is the best investment of time, focus and attention for leadership. For leadership to be successful in the art and science of developing people, there must be available methods, tools and techniques for coaching and mentoring, planning and preparing, training and educating, organizing and establishing knowledge transfer and competency groundwork.

People tools and techniques are a necessity for assisting leaders with their true role and responsibility of leading people. Over the past 50 years, I've been exposed to a plethora of tools and toolboxes targeted at supporting, developing and evolving people. During my time in academia, throughout my stint with NCR, AT&T and Bell Labs, and more recently for the past 30 years during my consulting business, I was exposed to, and experienced, a multitude of methods for training, improving and supporting people. Most of the People tools and techniques I've been exposed to are readily available and can be easily sourced and adopted. There are several that fit in well with the Affordability Paradigm, some of which are well known, and some of which are less well known.

SOME PEOPLE TOOLS TO CONSIDER FOR AFFORDABILITY

I've selected several methods and means that leaders may use to help and support the growth and development of people. Some of them I've been directly involved in defining and constructing, others I've adopted and adapted for application with my clients. I've found that all of them have been beneficial at one time or another

"The Progress Principle" (by Dr. Teresa Amabile and Dr. Steven Kramer)

In the spring of 1995, I was invited to participate in a Harvard Research Study led by Dr. Teresa Amabile focusing on events that motivated people in the workplace. Dr. Amabile assembled a team of seven field professionals, two research associates and two professors called the "IRG Team" (IRG: Independent Research Group). Research is normally executed by a professor and a research associate. From 1995 through 1999 the IRG Team led and directed by Dr. Amabile executed the study named The Harvard T.E.A.M. Study (i.e., **T**he **E**vents **A**nd **M**otivation).

Over a period of five years, the study encompassed seven companies across three industries involving 238 people from 26 project teams that averaged four months in duration. The project teams were primarily composed of professional knowledge workers whose work required them to solve complex problems creatively. The core function of the study collected input by way of a daily email questionnaire that queried study participants about events that happened each day within their "inner work lives." As a result, almost 12,000 responses (exactly 11,637) were gathered that journalled the daily experiences in the work life of the study contributors. It turned out, the most important query was: "Briefly describe one event from today that stands out in your mind." Beginning in 2000, Dr. Amabile and Dr. Kramer, with the assistance of various research associates over the next ten years, consumed all of the inputs, index coded, analyzed, synthesized and integrated all of the discoveries to create the book "The Progress Principle" (Published in 2011).

From the outset, it was never intended to study "Inner Work Life" (the conditions that foster positive emotions, strong internal motivation, and favorable perceptions of colleagues and the work itself). Positive inner work life starts with giving someone something meaningful to accomplish. It requires clear goals, autonomy, help and resources—what people need to make real progress. It depends on showing respect for ideas and the people who create them. The Progress Principle involves three intertwined dynamics of factors:

- The Progress Principle: The Power of Meaningful Accomplishment. Progress in meaningful work (small wins, breakthroughs, forward movement, goal completion)
- The Catalyst Factor: The Power of Project Support. Catalysts are events that directly help project work (setting clear goals, allowing autonomy, providing resources, providing sufficient time, helping with the work, learning from problems and successes, allowing ideas to flow).
- The Nourishment Factor: The Power of Interpersonal Support. Nourishers are interpersonal events that uplift the people doing work (respect, encouragement, emotional support, affiliation) all lead to a positive inner work life.
- Primacy: Progress in meaningful work
- Negative Forms (setbacks in the work, inhibitors—hinder projects, toxins—interpersonal events that undermine the people doing the work. Note: Negative events are more powerful than positive events all else being equal.
- Even seemingly mundane events—such as small wins and minor setbacks—can exert potent influence on inner work life.

In summary: Focus on Progress "Manage for Progress" (I'll add "Lead for Progress"). Provide meaningful work, provide leadership support, build strong teams and encourage teamwork, and keep in mind that "negative events are more powerful that the positive event." For more information on "The Progress Principle," read the book of the same name.

The Balance of Trust (by Dr. Stephen R Covey)

The Balance of Trust goes like this … Trust is like a bank balance. It operates as a "trust account" between two people. If there's a positive event it is considered a "deposit of trust." If there's a negative event that's a "withdrawal of trust." It's important to note, that one deposit does not equal one withdrawal. Withdrawals are ten times more powerful than deposits. To keep the "trust account" in balance it is required for ten deposits to be made to balance one withdrawal. Lesson: Leadership must continually deposit into the people's trust accounts, because somewhere along the lines of work and operation, withdrawals will likely be made. Keep depositing into the trust account of others.

The Seven Habits of Highly Effective People (by Dr. Stephen R. Covey)

Since many people are familiar with this tool, I'll summarize. The seven Habits help a person grow from Dependence to Independence to Interdependence. The seven Habits are:

1. **Be Proactive**: Take responsibility for your life and actions.
2. **Begin with the End in Mind**: Define clear goals and outcomes.
3. **Put First Things First**: Prioritize tasks based on importance rather than urgency.
4. **Think Win-Win**: Seek mutually beneficial solutions in relationships.
5. **Seek First to Understand, Then to Be Understood**: Practice empathetic listening.
6. **Synergize**: Collaborate effectively with others to achieve better results.
7. **Sharpen the Saw**: Engage in continuous self-improvement and renewal in all areas.

There is also an eighth Habit: **Finding Your Voice** and Inspiring others to find theirs. Your voice is a combination of combination of talent, passion and conscience. Find your voice then influence and encourage others.

Servant Leadership (by Robert K Greenleaf)

I learned about Servant Leadership while working with Bell Labs during the mid-1990's. The concept was created by Robert K Greenleaf in the late 1960's and was published in his 1970 essay "The Servant as Leader." His philosophy emphasizes that the primary role of a leader is to serve others. Within the Theory of Affordability, leadership serves those directly in the Value Stream, everyone that supports the Value Stream and all personnel of every function and department. Leadership must set direction, align resources, motivate people, communicate the message and execute the plan, as well as model the way, inspire a shared vision, challenge the process, enable others to act and encourage the heart in support of all the people of the organization. Greenleaf describes a servant leader using ten characteristics: listening, empathy, healing, awareness, persuasion, conceptualization, foresight, stewardship, commitment to the growth of people, and building community. For more information, see the works of Robert K Greenleaf on "Servant Leadership."

The 3 M's (by Dr. Rosabeth Kanter)

While participating on the Harvard research T.E.A.M. Study IRG Team, I was introduced to Dr. Rosebeth Kanter, a friend of Dr. Teresa Amabile. I learned of an interesting tool she created that related to worker motivation, competency and teamwork. She called it "The 3 M's." In a nutshell, the 3M's stood for Meaning, Mastery, Membership.

- Meaning refers to the purpose and fulfillment your job provides. The level of satisfaction with the job you have and it's role as an important contribution to the organization.

- Mastery is the ability to pursue excellence and expertise in your specific work area. It pertains to personal and professional growth.
- Membership is about belonging and connection with co-workers. Being part of a collaborative team provides motivation and establishes sense of purpose.

An individual's loyalty to work, and commitment to job, can be assessed using three simple 3M questions:

1. For the job you have, is it meaningful and does it have the same meaning you expected?
2. Are you encouraged and permitted to achieve the growth you need for your job?
3. Do you feel that you belong and are "part of the team"?

Note: If the answer to all three questions is "YES," the individual expresses job loyalty and will likely to continue to work at that job. If one or more answers are "NO," then beware, that individual may be seeking to change jobs.

THE 6 C SUCCESS CONTINUUM

I've used "The 6 C Success Continuum" on several occasions to evaluate the current state of a team, and discuss the possible future state and how to get there. Its also a simple, straight forward and effective instrument like the others that I use in my practice. I typically paste it up on a wall and describe details of each of the six categories (Note: The continuum line is a little more than six feet long with each category area of 12 inches.). Each participant is provided with a 1" round colored dot sticker (same color for everyone), and then asked to place their dot where they think the team exists today.

Coercion	Conflict	Co-Existence	Cooperation	Collaboration	Co-Ownership

Once all the dots are placed, I write above each category a descriptor that emphasizes the dynamic of each dimension.

"Chain Gang"	Tension	Cohabitate	Aid & Assist	Teamwork	Owned Results
Coercion	Conflict	Coexistence	Cooperation	Collaboration	Co-Ownership

A team dialogue and discussion is held to ask the "Why Questions" to get a collective understanding of the current state. Then, the discussion is shifted to "What can we do to improve?" Dots are again distributed (using a different colored dot) and each individual places their dot where they think the team can arrive over the next 6 to 12 months if improvement is pursued.

Afterward, the team decides actions and activities to execute to improve team performance, and the team lead documents, distributes, schedules and monitors the activities.

TEAMS AND TEAMWORK

For me, the ideal size of a team is 5 to 8 with 12 being the upper most limit in size. I prefer cross functional teams for problem solving, mixed discipline teams for product, service and process improvement and focused, subject specific teams for work groups and projects. My first experience with self-directed work teams was at NCR when the assembly area of the plant had a manager for first shift, a manager for second shift and eight self-directed, self-led work team comprised of more than 200 people each shift. From my experience with sports team from early on in my life, to various forms of work teams for the past 50 years, I am convinced through observation, research and experience, that well designed teams perform work better than separated individuals (Note: Cooperation and Collaboration lead to Teamwork and Co-Ownership … "more than two heads are better than one").

PICKING THE RIGHT PEOPLE

We most often pick people for their knowledge, expertise, experience and a proven track record, but to employ people that also demonstrate passion and enthusiasm for the job is a more difficult task. Traditionally, individuals were hired for their "heads" of their "hands," but the super achievers I've experienced had their heart and their soul in their work. A resume or vitae does not convey the individual's emotional compassion and spiritual nature toward a profession, a trade, or an occupation. Only through one-on-one discussions and dialogues can anyone truly assess the ability, capability and likelihood an individual can succeed at a job. There are times when the "perfect pick" emerges, and the candidate comes through with crystal clarity in terms of their fit, and all the pieces of the puzzle fall in place.

It reminds me of a short story from back in 1991 that helped shape my philosophy on hiring. I was the coach of my son's under six soccer team. Although most of the teams were comprised of boys, each team had at least two girls on the roster. One of our girls had six older brothers who all played soccer. Her obsession and desire was to play goal keeper. Her dad told me that she had no other interest in any other position because her brothers would play 3 on 3 soccer in the back yard and make her be the keeper. Most of all, she took pride in stopping the boys from scoring a goal. That year, we didn't have a single goal scored on us the entire season, and we went undefeated. The right players in the right positions create an enterprise of performance excellence. It's not just about "the hands and/or the head," but also "the heart and the soul."

Personally, I did not rely on solely on HR for my hiring needs. Over time, I learned that HR's priority was to support the company first and the individuals second. In fact, working with one client, when the VP of HR would come into the room of HR Employees at the beginning of a meeting and ask, "What is the role of HR?" … and they would reply in unison, "To Protect the Company," the pieces of the puzzle all clicked in place for me. And traditionally, in my experience, the role of HR has always been "to protect the company."

Jack Welch, when he was CEO of General Electric (GE) operated with 20-70-10 rule for organization personnel. The first group, the top 20%, were the high performers. The second group, the middle 70%, were the average employees. The third group, the bottom 10%, were the low performers that should be let go. He believed the top 20% should be significantly rewarded to motivate them further, while the middle 70% should be trained and supported to help them improve their performance, and the bottom 10% were to be removed in order to maintain a high-performance culture.

Spend the extra time upfront to not only choose and hire people who have the capability and intellect but find a way to employ the ones that also have the heart, passion, soul and spirit to perform above the norm. Skills can be developed, character can be influenced, but the desire to succeed, the appetite to accomplish and the belief in oneself is difficult to teach and instill.

WHEN ITS "GO TIME"

Nick Saban, a famous football coach, has a notable quote that emphasizes the importance of the player over "the play": "*It's not the play; it's the player.*" This quote reflects Saban's belief that the quality of a player's performance is more crucial than the execution of a play itself. Recently, during the National College Football Championship, the Indiana Quarterback, Fernando Mendoza was faced with a 4th Quarter, 4th down and 5 yards to go on the 13-yard line. There was 9:27 left in a game, and the coach called a Quarterback Draw Play that put ball in Fernando's hands to either make a first down, score a touchdown, or fail and put the ball back in the opponent's hands. Long story short, after a great effort, Fernando scored a touchdown (Note: its well worth watching any YouTube Video of the play.). That touchdown was the game winning score, Miami 21—Indiana 27. Even more interesting is the back story that led up to that Championship Game.

Prior to 2025, Indiana sported the 'loosingest' record in College Football history, and went from the worst team in College Football to be the undefeated National Champion in the 2025–2026 season. Even better, Fernando Mendoza, their Heisman Trophy Quarterback, was rated a 2-Star quarterback out of high school, ignored by 134 colleges, rejected as a walk-on of his favorite team Miami, playing third string at California before transferring to Indiana to eventually score the touchdown that won the game. In summary, the coach believed, the team believed and Fernando always believed. Learning: Heart and Soul trumps Mind and Body, "When its go time," it's the player, not the play.

STAFFING AND SCHEDULING TO CUSTOMER REQUIREMENTS AND COMMUNITY SOCIOECONOMICS

Sometimes, in staffing, hiring and scheduling people for an organization, special conditions in the community may exist that can be leveraged as an advantage for the company. About 35 years ago, while employed by NCR, I was exposed to such a situation. In Duluth Georgia NCR operated its Just-In-Time Retail Systems assembly plant. It was chartered to deliver NCR systems to Customers in the Retail business around the world. In American Retail, the highest sales months are November and December

("Shopping Season"), followed by August/September ("Back to School") then March/April ("Spring Fashion") with the rest of the times January/February and May–July rounding out the revenue flow. Our Customers did not want us to install equipment at the "high time" (October, November, December) and just after the holidays because of yearly inventory, and they preferred their installs to ramp up in frequency in the spring and begin ramping down at the end of the summer. Our demand was low at the beginning of the year, ramping up into the summer, and beginning to ramp down as the end of the year approached (Note: Almost like a 'bell curve').

Staffing to demand and scheduling of resources for assembly required heavy staffing during the part of the year with high demand and light staffing in parts of the year with low demand. We knew that cyclical layoffs of workers was de-motivational, so we sought a formula that would employ a set number people on a permanent basis throughout the year, with the ability to increase our population with "permanent flexible" resources when demand ramped up ("Permanent Flexible"—people who are willing to work on a part-time basis such that they didn't work during slow time, and were committed to work during high demand times).

Fortunately, we had a substantial portion of our community represented by people of Asian heritage and culture, predominately Korean. Many of the families were multigenerational where parents placed a priority on childhood education. We found that we could staff our "permanent flexible" workforce with men and women during high demand for manufactured product (i.e., late spring and summer) who had grandparents at home to watch-over the children, and permit many of these employees to be at home during most of the school months to tend to their children's educational needs.

Lesson Learned: Creative Thinking and Complex Problem Solving combined with Social, Cultural and Economic Conditions can serve to resolve issues and challenges with difficult personnel matters.

WHEN IT'S TIME TO FLUSH THE TOXICITY

Full disclosure in the spirit of truth, trust and integrity, not all the efforts I've been engaged in over the past 30 years have been spectacular sunsets, rainbows and unicorns. When I'm asked, "What have been the greatest, most difficult challenges you've encountered?" I answer, "Leaders that resist and undermine advantageous change, transformation and improvement, by either overt means or covert methods, with vindictive actions and vitriolic attacks." One of my favorite examples comes out of Conroe Texas from a project where I was invited in to serve the merger of Byrne Medical (Conroe Texas) and Medivators (Minneapolis Minnesota), by implementing and incorporating Lean Methodologies and Practices within their Manufacturing Systems.

Most of my time with this project was spent with manufacturing work teams transferring Lean knowledge, certifying Lean Green Belts using an accredited IISE program, and coaching/mentoring workers during their certification projects. By design, I was required to deliver a one-week leadership session intended to expose and engage 24 top Leaders of both organizations to the Lean Philosophy and the actions, activities and projects of the workforce. The room was full of SVPs, VPs,

AVPs and Directors representing engineering, manufacturing, strategic marketing and sales, finance and purchasing, customer support and top management. There were 12 tables arranged in four rows (i.e., three across and four deep), with two people at each table. The highest ranking VPs from both companies sat in the back row. We started at 8:00AM, broke for a lunch hour at 12:00PM (Note: Excellent Lunches every Day) and ended at 5:00PM on the dot.

The first day, Monday, was mostly an introductory day with team exercises in the morning and afternoon, and ended on a high note of support and acknowledgement delivered by the top leaders in the back. Tuesday was much the same containing a bit more discussion and a good "walk about" through the manufacturing part of the facility. The dialogue and discussion increased on Wednesday and Thursday, when the hand-full of advocates (i.e., five) in the room were easy to identify, the 15 wait-and-see/middle-of-the-roaders became apparent, and the few (i.e., four) resistors began to emerge. Its worth mentioning, it was easy to observe who had clandestine meetings during the breaks and who at lunch at a table separate from the others.

On Friday about mid-morning, the devil "came out of the details," and all hell broke loose. The main and highest-ranking inhibitor, the VP of Marketing and Sales for Byrne Medical (sitting in the middle table on the second row) began to challenge, argue and dispute about every topic I covered. I employed a technique I learned from Dr. John Kotter that he calls, "a counterintuitive strategy for saving your good idea" (see the book "Buy*In," pgs. 87–103). It is architected as follows: Work to achieve Buy-In by 1. Capturing Peoples' Attention, 2. Then with people paying attention, winning over their minds with data, facts and truth, 3. And with people paying attention, also winning over their hearts. During Monday through Thursday, I realized I already had Buy-In from 20 of the 24 people in the room. I treated the top resistor (and his three compatriots) with respect and dignity, but didn't compromise my position or situation. I countered his attacks with more data, facts and truth, while mostly paying attention to the other 20 people in the room, reading their reactions and observing their nods of support and verbal responses of rebuttal to the confrontations during the morning topics. As we broke for lunch, the SVP of Engineering and Manufacturing from Medivators, caught my "adversary" in the hallway, and for lack of a better term, "read him the riot act" (Note: according to my Texas advocates, he "took him out back of the wood shed"). Everyone in the room heard the SVP dressing down the VP before both returned for lunch. The rest of the day went quite well, and we ended on a very positive note.

After about a year, both facilities accomplished and achieved a great deal of process improvement and cost reduction. And, its worthy to note, my VP challenger was no longer with the company. My confidants told me, "He was flushed."

YOUR PATHWAY AND YOUR JOURNEY

Recently, I watched two American Hockey Teams win gold medals at the 2026 Winter Olympics. The men's team won gold in 1960, 1980 and 2026, while the women's team won gold in 1998, 2018 and 2026. Both teams won with a score of 2–1 in overtime against their rival Canada. As I looked back after the victories, I asked myself, "What was it that both Americans teams had that enabled them to beat their Canadian rivals?" I found it best expressed by Greg Wyshynski of ESPN in his February 19, 2026 article: "Why the U.S. women won—and Canada lost—2026 Olympic gold." … He said,

> Being the best in the world is about finding a way to win when things aren't clicking and the opponent isn't relenting. When there's enough pressure to turn a lump of coal into a diamond—which is actually a pretty good description of how the Americans won the gold medal over Canada.

Leaders must develop a workforce that can "find a way when things aren't clicking and the opponent isn't relenting. When there's enough pressure to turn a lump of coal into a diamond"!

> Mine the mountains of candidates out there, gather the gold nuggets, then teach, coach and mentor them so that they can turn coal into diamonds.

10 Process

> If you can't describe what you are doing as a process, you don't know what you're doing.
>
> **– Dr. W Edwards Deming**

PROCESSES FOR AFFORDABILITY

All work is a process. A job or assignment to perform ongoing work with no beginning nor ending requires the use and utilization of a process and is likely to involve several processes to complete the assigned tasks. Work that is framed by time with a beginning and an end, better known as a 'project', involves a project management process as well as other processes and procedures to complete the project. Whether the processes are well documented or "tribal knowledge" (i.e., passed from person to person verbally without any written words or drawn images), the means and methods to perform work requires the knowledge of processes, procedures and tasks. In smaller organizations, it is commonplace for formal documentation, training and management for processes are absent and weak at best. Larger organization are typically relegated to formalize work processes with documentation, training and management to meet knowledge and capability demands of the workforce.

THREE VIEWS OF A PROCESS

The architecture of a process, in its simplest form, is referred to as a S.I.P.O.C. (i.e., Suppliers, Input, Process, Output, Customer). Suppliers provide Input(s) for the Work Process to transform Input to create Output(s) that is delivered to the Customer (Note: "Customers" can be either external or internal or both to the organization.). When creating or documenting a process the five SIPOC elements should be described in detail. Every supplier and all specific inputs should be clearly declared. Each step of the process, in the sequence of occurrence, should be listed. The results or outcomes of the process should also be identified. Finally the customer or customers should be listed (Figure 10.1).

Here is an example of a SIPOC for the Process of Preparing a Program for implementation.

Program Preparation Process S.I.P.O.C.

As you can see, the suppliers for the Program Preparation Process include the Customer, participants the customer has designated for participation, partners and colleagues of the customer, S.M.E.s (Subject Matter Experts) and the facilitator for the program. Input to the process includes all information and intelligence needed to prepare the program. The seven steps of the process are detailed in the sequence of occurrence, and the results of the process are listed in the output column. Lastly, program participants make up the customer list.

DOI: 10.1201/9781003772934-10

Program Preparation Process
S.I.P.O.C.

Supplier	Input	Process	Output	Customer
Customer	Customer Expectations	1. Arrange for a Meeting/Discussion	Project Purpose/Intent	Customer
Customer Designees	Project Requirements, Needs, Wants, Wishes	2. Determine Project Content/Details	Deliverables	Facilitator
Partners/Colleagues	Goals, Objectives, Measurement(s)	3. Define Timeframe and Logistics	Content	Designated Participants
S.M.E.s	Current State and Baseline Information	4. Identify Participants	Timeframe(s)	
Facilitator		5. Build a Plan for the Assessment	Logistics	
		6. Declare the Elements of Success	Participants	
		7. Document the Project	Project Doc. and Assessment Plan	

FIGURE 10.1 A S.I.P.O.C.

A SIPOC can be used to further document the process using text, graphics, illustrations and drawings. Here is a flowchart illustrating the Program Preparation Process (Figure 10.2).

Processes are best documented using visual tools often supported by written documentation. Visuals have a better effect when training and developing the capabilities of the workforce. Diagrams, drawings, pictures and videos add a better level of understanding and transfer knowledge more efficiently.

In this particular case, a flowchart was used to illustrate the process. This flowchart was created using Microsoft's Visio. There are numerous process documentation tools available on the market. I suggest some thorough research and evaluation be done prior to purchase. Warning: Be careful, evaluate several judiciously, be aware many are not comprehensive.

VALUE STREAM MAPPING

For an Enterprise System-level view of processes, I strongly recommend Value Stream Mapping. Value Stream Mapping provides a "30,000 foot view" for the Enterprise (Note: It can be adopted at lower levels when its necessary to look at Sud-Streams of Value and Value Support Streams.). It serves as a System SIPOC for illustrating the flow of Value, and can be utilized to identify major area of waste.

Here is an actual example of a simple Value Stream Map for a Patient's Physician Visit that illustrates how Value is Delivered, how the process flows, providing a documented record of the Enterprise "SIPOC" with metrics and measures, as well as the identification of waste for targeting improvement events and activities (Figure 10.3).

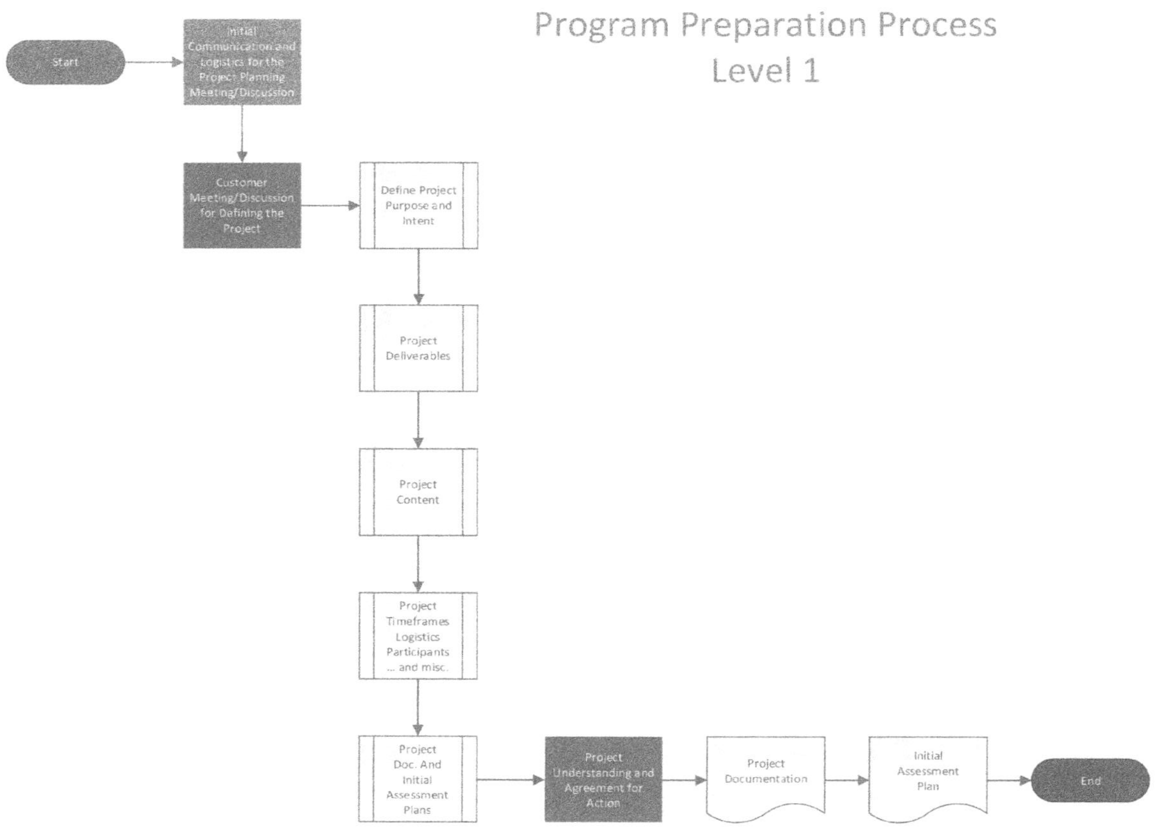

FIGURE 10.2 A process flow chart.

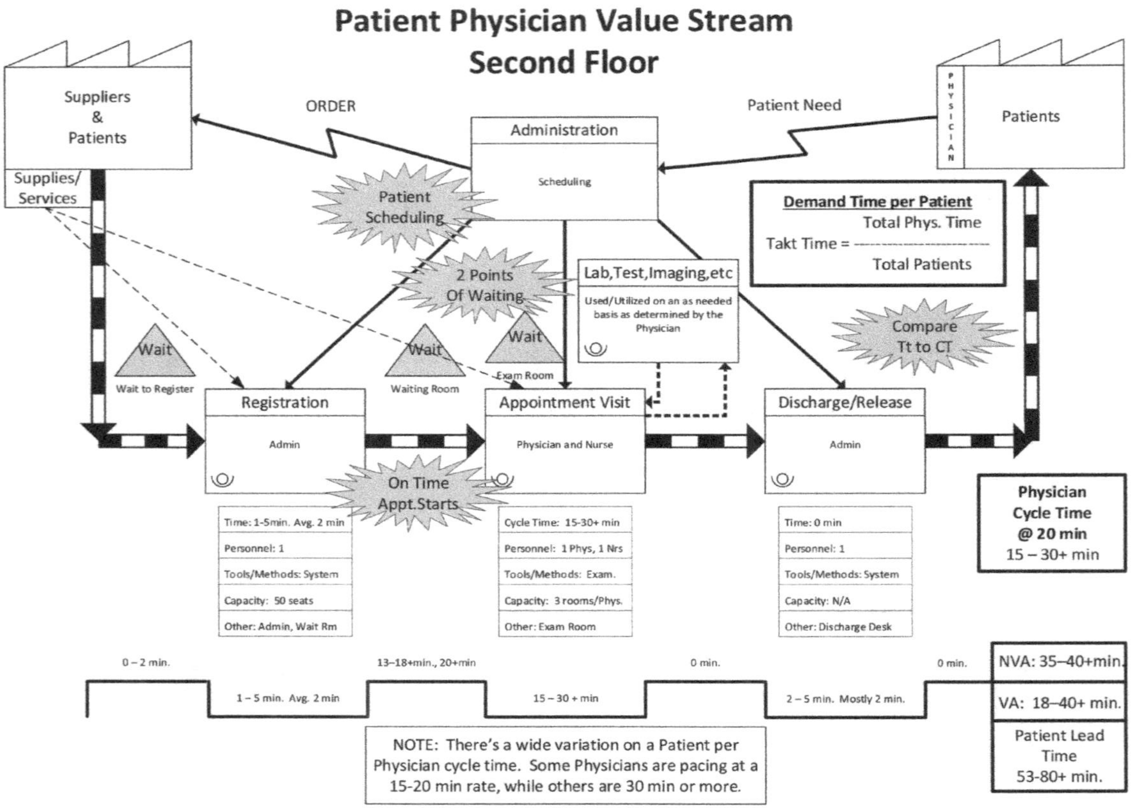

FIGURE 10.3 An example of a value stream map.

I've only presented three different, yet interrelated, examples of process documentation. With technology today, process documentation can be online (Note: Permitting virtual access), can include videos (Note: Allowing workers to "see how its done"), and can be created by way of the use of a multitude of available tools.

Process documentation is only the first step for process maturity. In order for a process to be efficiently and effectively managed, the process must be well documented, stabilized, standardized and predictable in terms of performance. If you were to apply an advanced tool, such as Six Sigma for process improvement, all four factors must be in place. After documentation, you can begin to stabilize a process by eliminating waste. Standardizing a process ensures its consistency throughout the enterprise. Predictability occurs when the process performance is consistent and outcomes can be easily forecasted.

Managing a process involves constancy of purpose and the commitment of continuous improvement. Dr. Deming's process quote is not only valid, but true in every case I've experienced in the last 40 years. If a process is transferred by way of mouth, that's tribal knowledge which tends to be different every time its communicated. To begin process management the steps are simple: 1. Document, 2. Stabilize, 3. Standardize, 4. Measure for Predictability, 5. Manage by way of continuous improvement (Note: Be sure to keep in mind, that even if the five steps of process management are working, there might be a better version or different version of the process that yields greater performance.). Always be in search of a better way of doing things. Either doing things better with the process you've got or doing things better with a different process that may incorporate different means, methods and techniques as well as new technologies and tools.

PROCESS HIERARCHY

At the Enterprise level, there are systems in place comprised of processes. As an example, a Value Stream (i.e., A System) is comprised of supplier input, value creation, value production and value delivery (i.e., All Processes) that output (i.e., Value) to the Customer (Figure 10.4).

Note: The concept of a Value Stream emphasizes Flow. Flow moves 'upstream' inputs to 'downstream' Customer(s). The Goal is for the Value Stream to Flow as smoothly and effortlessly as possible. Influences such as Customer Demand Pulls the Flow. Waste, defects, roadblocks and barriers impeded the Flow of the Value Stream. Affordability uses the Value Stream concept for Value Delivery, with Flow and Pull, with an ever-vigilant Goal of the Elimination of Waste (Figure 10.5).

At the System Level, processes may take on different forms. Here is a sample of a process that illustrates multiple flows.

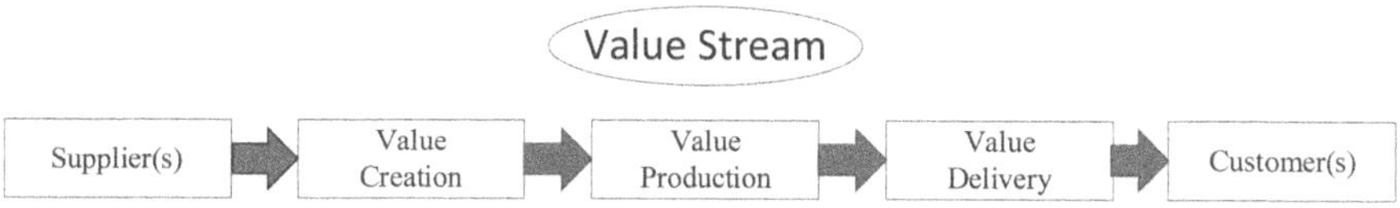

FIGURE 10.4 Value stream basic elements.

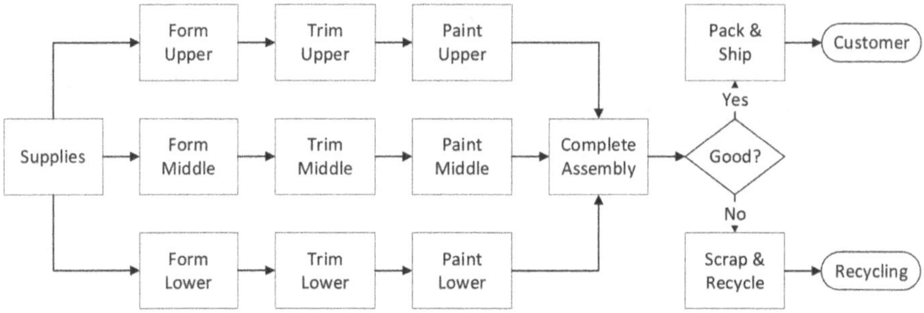

FIGURE 10.5 A process flow illustrating parallel processes.

Processes can be linked and linear (like the example above), processes can run parallel to each other, processes can be merged or diverged, or processes can occur in a multitude of flow forms or shapes. Processes are the actions and steps within a system (Figure 10.6). The chart below diagrams the hierarchy of an organization from an enterprise level to the team/individual level, and depicts where systems, processes, procedures and tasks fit within a system (Note: An entire organization is composed of Systems that define all of the innerworkings and functions of the institution.).

Process documentation containing process maps, diagrams and illustrations can be used as a standards policy, a reference for workers, training material, even as a sales tool to explain to a customer how value is created, produced and delivered. Overall, process information should be used as the center of knowledge for all work (Note: It has been proven, that visual process documentation is much more effective and efficient than textual forms. Use diagrams, drawings, pictures, videos and maps wherever possible.).

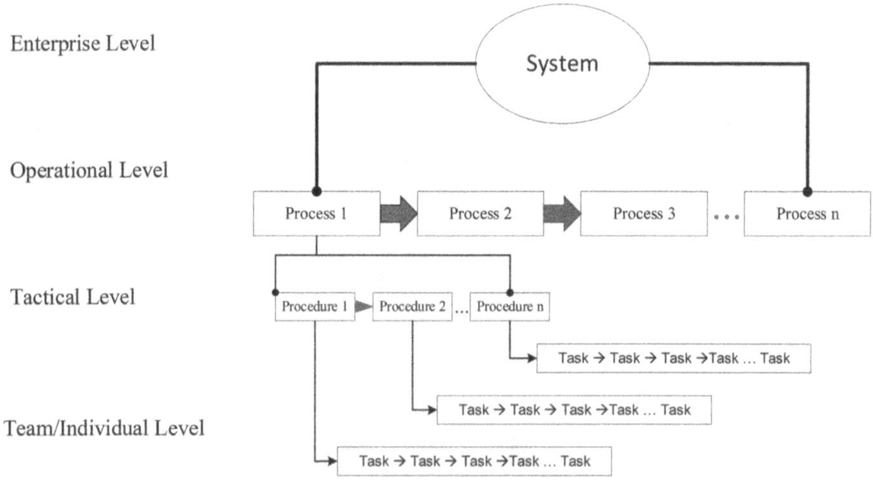

FIGURE 10.6 The hierarchy of an Enterprise including the levels of work.

BUILDING AND MAINTAINING A PROCESS ENVIRONMENT

There are several steps required to build and maintain a process environment. As always, every path that leads to Affordability begins with the first step of understanding. The stages from creating to sustaining a process are as follows:

1. Understanding: Observe, study, research, investigate and get to know process details.
2. Document: Record, illustrate, map the details of the process.
3. Stabilize: Operate the process, understand performance, eliminate waste and variation.
4. Standardize: Normalize and systematize the process to make it predictable.
5. Maintain: Monitor performance and make adjustments as necessary.
6. Sustain: Seek and attain performance excellence, continuously improve as necessary.

Please note that there are several formidable tools and toolboxes available for the continuous improvement of processes: Basic and Advanced Quality Tools, Lean Thinking and Principles, Six Sigma Tools and Techniques, as well as others in the marketplace.

A CASE OF THE 'IF IT WORKS THEN DON'T FIX IT' FALLACY

About a dozen years ago, I was invited in to deliver the accredited Lean Black Belt Certification Program of the IISE for the Health Division of the architectural firm Perkins&Will (Note: A Top Ten Architect Worldwide, AKA: P&W or P+W). The sponsors for the program Richard Herring (a Principal of the firm) and Marvina Williams (a Registered Nurse and Expert in Healthcare Design), wanted to certify Lean Black Belts throughout the Health Division in order to meet emerging Customer requirements that included architectural expertise with Lean Black Belt capability. This was a newly emerging initiative in the Healthcare Marketplace. Programs were delivered in Atlanta (the Health Division's Main Office), Washington DC, Raleigh, Dallas, Chicago, Los Angeles and Seattle. The primary purpose was to deploy Lean Knowledge, Lean Thinking and Lean Practices throughout the P&W Health Care Architects and within Health Division Designs of Perkins&Will (Note: Some architects from other divisions participated.).

Each participating architect had to pass an exam and complete a Lean Project (Figure 10.7). Within the curriculum, the most studied and discussed process was the commonly used and taught Architectural 7 Step Process: 1. Pre-Design, 2. Schematic Design, 3. Design Development, 4. Construction Documents, 5. Bidding and Negotiation, 6. Construction, 7. Project Closure and Occupancy.

FIGURE 10.7 The traditional architecture process.

This process has been taught in Architectural Schools for the past 70 years. Its well documented, stable, standardized and predictable. However, as part of the Lean Black Belt Program, Richard directed the Atlanta Office to re-engineer the process and create a new process that was more efficient, more effective, faster, better and more Affordable to execute. The new process (which remains proprietary as a competitive advantage to P&W) has a somewhat similar flow, where some of the elements and facets have changed, some of the process steps have been moved earlier, and the customer involvement and engagement is from the concept of thought or fragment of an idea, to beyond the project closure and occupancy.

Today, Perkins and Will remains to be a premier architectural firm serving 20 different target markets worldwide on the leading edge of research, innovation and implementation.

A PROCESS TO PROTECT LIVES

Earlier in this book, I made mention of the MRAP Program (2006–2014) that deployed more than 27,500 mine resistant vehicles to Iraq and Afghanistan in three years. During and after the deployment, the vehicles on the battlefield were exposed to clashes and skirmishes with the enemy that sometimes resulted in damage (e.g., wheels, fenders, doors, undercarriage, engines, etc.). Each country contained several camps and forts that had repair and service centers for the trucks. By 2011, every one of the repair depots had a large backlog of damaged vehicles. The backlog situation impacted on the rhythm of the war since it delayed units from returning to combat.

The MRAP Program was called upon by the Pentagon and Central Command to address the problem. Battle Damage And Repair (AKA: BDAR) quickly became a top priority. The Program Manager, Dave Hansen, assembled a team of 5 individuals for the purpose of addressing the 'crisis'. The team was comprised of a USMC Colonel (Team Lead), three MRAP Program colleagues and me. I served as a Lean Six Sigma Master Black Belt expert on the team. The Goal was to increase the speed, improve the quality at an Affordable cost for BDAR activities. After training and preparation for battlefield exposure, we were first deployed to Camp Leatherneck in the Helmand Province of Afghanistan during the beginning of June 2011. We flew from Atlanta, to Dubai, then over Iraq, to the US airbase in Kandahar and then to the Camp Bastion British airfield that was attached to Camp Leatherneck.

When we arrived, we met with the camp commander and I soon found out, our team lead, Col. Brian Fulks was previously the USMC officer in charge of BDAR Operations years before when the use and utilization of MRAPs just began. After our meeting we toured the area and the three facilities where the different contractors repaired the vehicles (Note: The repair building were known as K-Spans). To put it in a few words: "the place was a mess, quite chaotic and the BDAR processes were definitely not standardized." That day, I was able to take a lot of notes, get a good baseline understanding established, diagram the current state conditions and determine the operating metrics and measures.

The "yard" outside each of the three K-Spans, was full of trucks waiting to be serviced. There was no "rhyme nor reason" for how the trucks were organized. Additionally, it wasn't clear as to how the inventory of trucks were prioritized for service. Also outside the K-Span, were piles of materials and dozens of conex containers housing the parts for service. Inside each K-Span there were five to six trucks receiving BDAR services and several teams of different contractors using different processes for the repair work. When a truck capsule needed to be removed from the truck body, a mobile crane was used that caused all work to cease for safety reasons. Because of the inconsistent practices within the K-Spans, there was a lot of wasted space.

The first step was to organize the workspace inside and outside the K-Spans. Inside the K-Spans a standard approach to configuring service bays was concocted using an angular parking design that increased the capability from five to six trucks at-a-time to 10 to 12 trucks at-a-time. Outside the K-Span and in the truck yard, the parts pile clutter was eliminated, the parts flow was organized to go directly from a conex container to the repair bay, and the truck inventory was re-organized to flow from a specific parking space to a K-Span for repair in priority order as set forth and communicated by an order board in front of each K-Span.

The BDAR processes used were re-engineered to follow a standard approach to repair. The entire BDAR system and processes originally operated on an average of nearly 20 days to repair. The rate of repair was cut in more than half the time after the processes, procedures and tasks were standardized. A young USMC welder invented a "capsule cart" that he created from angle iron steel and heavy-duty wheels where a truck capsule could be placed and serviced in a flexible and mobile manner (Note: This reduced the demand of crane use and eliminated the numerous work stoppages for safety.).

All in all, with the K-Span capacity doubled, the time-to-repair rate cut in at least half (i.e., on average a week or less), and the volume of trucks serviced in each K-Span was four times faster than the baseline. The BDAR Processes operated faster, the repair quality was better, and there was less waste of parts, materials and time for the contractors who obviously increased in pride-of-workmanship and personal satisfaction of a job well done (AKA: "Affordability"!). What I also found out while I was there was the reality that USMC units refused to use a Humvee in place of their MRAP that was waiting to be serviced.

With the proof of concept complete, and the system standardized, the same Affordability Design was replicated at the BDAR sites in Camp Kandahar and the Bagram Military Base by the end of 2011 (Note: The War in Iraq was ended by President Obama in December 2011 so the demand for the improved BDAR Design was not required in Iraq.).

THE TRUE VALUE OF A PROCESS

Affordability is based on the Value Stream Delivering Value. The Value Stream is an Enterprise System composed of integrated processes of a Strategic SIPOC, from suppliers to customers. All of the work, of all of the processes, both Value Add processes and Value Added Support processes, culminate in the Delivery of Value and ongoing Customer support. Processes focus the workforce on the work at hand. Processes serve to introduce, educate and direct how work is done and Value is Delivered. Know your processes. "If you don't know your processes, you don't know what you're doing."

YOUR PATHWAY AND YOUR JOURNEY

If you haven't already, take some time to completely understand the processes you use to deliver Value. Take the time to understand the processes you use to support the creation, production and the delivery of Value. The initial goal is to gain a comprehensive understanding of the current state of your systems and processes.

If you haven't documented your processes, document them. If your processes aren't stabilized and standardized, stabilize and standardize them. Measure their performance and compare the current state to the baseline and the goals and objectives set for performance. Identify the waste and variance that can be removed.

Define additional improvement opportunities (i.e., elimination of waste, variance, etc.). Using the current state and knowing the opportunities for improvement, render a future state for you to pursue. Develop a plan for improvement and always include the basic Affordability quantitative and qualitative performance measures: Time, Quality, Cost, Customer Satisfaction, Employee Satisfaction, Supplier Satisfaction. Execute the improvement plan and improve the process(es). The steps are simple: 1. Current State, 2. Identify Waste and Improvement Opportunities, 3. Craft a Future State and Develop an Improvement Plan, 4. Execute the Improvement Plan (Note: When complete, go back to #1).

Remember the 5 P Model. Purpose: People → Process → Performance → Profitability. Keep this in mind to include all dimensions and dynamics in your processes and systems. Use the S.I.P.O.C. Model for the basic architecture of each process. Flowchart your processes for the purpose of creating visuals for the people. Use Value Stream Mapping to diagram the Value Stream.

The documentation of processes and systems are key. Please remember, documentation is not only textual, but also illustrative and graphical to the extent of even using videos and digital imaging for communication. Affordability emphasizes that visuals are most effective to convey information, a message or a communique. What is more important is the stabilization, standardization, predictability and use of processes and systems by the people. Ultimately, it is the performance of the processes and systems that determine success of the enterprise. Know what you're doing, know your process … knowing your process, you know what you're doing!

11 Performance

Tell me how you measure me, and I will tell you how I will behave.

– Eliyahu Goalratt

PERFORMANCE EXCELLENCE FOR AFFORDABILITY

Within Affordability, performance is measured by the results of processes and people involved with the Value Stream and Value Stream Support. As Value is delivered to the Customer, critical measures determine performance (i.e., Ontime, Quality, Cost). Within every process of Affordability time, quality and cost are primary quantitative metrics to be measured. Considering the qualitative dimension of Affordability Performance customer satisfaction, employee satisfaction and supplier (or partner) satisfaction are the primary qualitative metrics.

Performance excellence occurs when all of the measures consistently meet and exceed their targeted goals and objectives, and those goals and objectives are at or above the competition. At the point where goals and objectives are consistently met, an assessment of the goals and assessments should be performed, and an adjustment should be considered. Performance excellence continues to happen as goals and objectives are continually met and exceeded, and adjustments are made in an effort to increase performance.

As an example, many good organizations operate around a 4-sigma (or 4σ) range of quality performance in terms of "defects per million opportunities" (i.e., 4σ = 6,210 defects per million opportunities, AKA: DPMO). If the organization has a quality goal of 4σ and consistently performs at this level or 'better' (i.e., less than 6,210 DPMO), then it should be determined if the organization wants to change the process to perform at 4.38σ (i.e., 2000 DPMO, which equates to 99.800% Quality) or 5σ (i.e., 233 DPMO, that is 99.9767% Quality). If a new process performance goal is set at 4.38σ or 5σ, then a process improvement project should be designed and implemented to increase performance of the process.

All too often there are too many metrics to measure performance of the systems and processes of the Enterprise. It is also true that there exist organizations that measure too few performance goals and objectives (e.g., If an organization focuses on money and money alone, this monocular view would destroy the company.). Considering the various structural levels of an organization (i.e., Strategic, Operational, Tactical, Team/People) in conjunction with layers of the Affordability composition (Value Stream, Value Stream Support, Leadership) a custom version of the Kaplan and Norton "Balanced Scorecard" (i.e., Customer, Business, Process, People) fits nicely at the Enterprise, System and Strategic tier (i.e., "Top Tier"). Operationally and Tactically basic quantitative metrics involve time, quality and cost, while qualitative measures involve customer satisfaction, employee/associate

DOI: 10.1201/9781003772934-11

satisfaction and supplier/partner satisfaction. For Affordability, from the top tier to the base layer, all metrics and measures should have some linkage and alignment.

Money is a good universal variable to use as a metric in terms of Affordability Performance. Affordability itself involves money in the "Cost" realm of Value-Customer-Cost Affordability Model. Remember, cost and the cost of money for the Customer is set by the price paid for a product and/or service, and the cost involving money for Value Delivery is for the means and resources required to create, produce and deliver the Value. Revenue is generated from Customers purchasing Value, and expense or spend is a result of resources required to Deliver Value. When revenue exceeds expense profitability is realized. At the Enterprise level, the Business category of the Scorecard would measure and monitor revenue, expense, spend and profitability. At the "base layer" level, cost in terms of expense and spend would be measured and monitored against a team or individual budget. The intermediate layers of operational and tactical would also be linked and require monitoring of their performance in terms of money. In Affordability, money is most closely associated with Cost, both for the Customer and for the delivery of Value. Prosperity and Profitability are accomplished when revenue exceeds expense.

Other than money, or "costs," time and quality are primary Affordability performance indicators, as well as qualitative gages of satisfaction for customers, people (i.e., employees or associates) and suppliers or partners. Customers, Processes and People are all part of a strategic scorecard that can be linked all the way to the base layer of teams and individuals that drive processes that result in performance (Note: see the 5 P Model for more information) through the operational and tactical layers.

Variants of time performance to consider for measuring include: on-time, lead-time, downtime, uptime, setup time, cycle time, process time, maintenance time, changeover time, wait time, put-away time, inventory time, pick time, package time, load time, delivery time, installation time, preparation time, production time, order-to-delivery time, order-to-cash time as well as many other time based metrics (Note: Time Metrics are not limited to the many mentioned!). Time can be translated to 'speed', time can also be expressed in terms of availability or accessibility, and time can be considered for length of time in systems, processes, procedures and tasks.

Quality is a category of performance within Affordability in the areas meeting and exceeding customer requirements, meeting and/or exceeding design criteria, reaching an aim of perfection, attaining a specific rate of defects (Note: as in DPMO), and performing at a specific level regarding process, product or service. The definition of quality (*'according to Webster'*) is "a particular standard of something as measured against other things of a similar kind; the degree of excellence of something, or a distinctive attribute or characteristic possessed by someone or something." Considering both definitions, in terms of performance, Affordability Quality is defined and determined by the Value and the Customer of an Enterprise.

The Performance Measurements for Satisfaction are encapsulated by the qualitative perception of the Customer, the People in the organization, and the Suppliers and Partners involved with the Value Stream. The quantitative measures are tangible and objective while the qualitative measures are intangible opinions and subjective responses. An Affordable environment takes advantage of qualitative information and responds by improving the Value, Value Stream and Value Stream Support within the Enterprise. Qualitative conditions directly impact loyalty of customers, employees and associates, and suppliers and partners. Loyalty, consistency and predictability are all human factors tenets of Affordability.

There are two types of metrics to consider when thinking about predictive and resultant measures. One type is in terms of a leading indicator that can be measured, monitored and used to predict likely outcomes. The other is a lagging indicator that provides information "after the fact" that reports what had happened and to chronicle results. A parts shortage is a leading indicator that predicts an impact on production. Money is a lagging indicator that conveys the results of customer purchases. Variants of time, quality and cost can serve as leading and lagging indicators (e.g., On-time, if predictive can be a leading indicator, and if impacted, can be a lagging indicator. Quality, at a level of performance excellence can be a leading market indicator of potential growth, and when defect ridden, is a lagging indicator forecasting a reduction of revenue. Cost in terms of a reduction may indicate more sales, but when increased by suppliers will impact expense performance.).

Once the right metrics are determined, the right amount of the target(s) must be defined so that the organization can aim. For this to be determined, a baseline must be known (i.e., the current state performance), a goal must be established, and an interim objective must be ascertained (i.e., the future state performance). The difficulty is in the 'How To'. How To get from 'A' (the baseline, the current state) to 'B' (the objective, the interim state) eventually to 'C' (the goal, the future state). The answer is within the practice of continual improvement or continuous improvement. This can only be accomplished by keeping score and responding to needs and requirements of the purpose, people, process, performance and profitability conditions.

KEEPING SCORE

A question I frequently ask throughout the organizations I serve, "How do you know if you're succeeding or failing, or winning or losing if you don't keep score?" In Chapter 2: The Goal, I shared a story, "***When Just-in-Time Wasn't***." As you may recall, it was about the pursuit of the Goal of being Just-in-Time for the Customer. In that case, it took the better part of five years to approach the goal. From a baseline of 60.2% (Note: Which equated to the on-time delivery of Value of less than 2/3rds of the time) to a five-year result of 92.0% (Note: Still in pursuit the visionary goal of 100% on time). A key factor of that achievement had to do with keeping score.

After 1990, the first full year of production and delivery, when 'the dust cleared', the result was 60.2% on time. Reporting and communicating the on-time performance was mostly internal to the assembly department and the assembly teams. During 1991, the posting of performance on bulletin boards was an initiative across the plant. There were eight processes and process teams that comprised the Value Stream (i.e., Receiving, Customer Order Processing, Surface Mount Technology

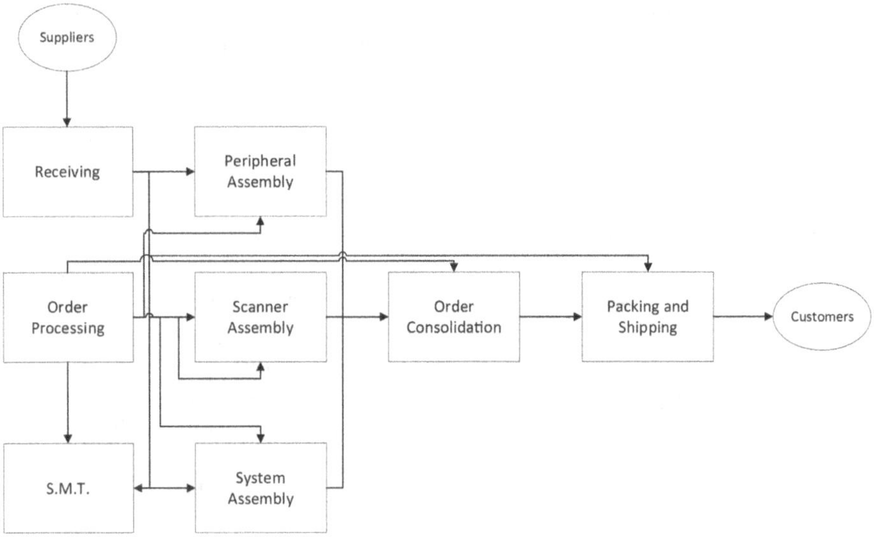

FIGURE 11.1 The system flow of the assembly operation.

AKA S.M.T., System Assembly, Scanner Assembly, Peripherals Assembly, Order Consolidation, Packing and Shipping) and by the end of the year, each team was posting their on-time delivery metric in the team's area, with the plant's on time measure clearly posted in each location.

Each team had "suppliers" and "customers." The receiving team had the incoming material from suppliers (i.e., their input), and their process involved the delivery and distribution of that material to their "internal customers" (i.e., SMT, System Assembly, Scanner Assembly, Peripheral Assembly, Pack and Ship). This Customer/Supplier SIPOC relationship was established throughout the Assembly area (e.g., Internal Suppliers: Order Processing, Receiving and SMT provided input to the Process: System Assembly who produced output for their "Internal Customer": Order Consolidation. It was an internal S.I.P.O.C.) (Figure 11.1).

Each team managed their own performance display board at their physical location in the plant. The "display board" consisted of three panels (3 – 4' × 8' panels): A People Panel (for posting information pertaining to the people on the team, the job requirements and responsibilities of the team and other pieces of miscellaneous team information), A Process Panel (containing a map of where the team was in the Value Stream, a Process Map and process details), A Performance Panel (Displaying Quantitative Metrics: On Time, Quality, Cost and Qualitative Metrics: Internal Supplier Satisfaction, Team and Associate Satisfaction, Internal Customer Satisfaction. Opportunities for Improvement, Improvement Projects with Status, Results in terms of Achievements, Awards and Accomplishments.) (Figure 11.2).

As performance improved and new objectives were set, during the third year, 1992, we started an initiative called: "Customer Plant Tours." This entailed customers, coming to the plant, and after an introduction and update, the visiting customers

People	**Process**	**Performance**
Team and Individual Portion	Value Stream Map & The Team's Process	Quantitative Time, Quality, Cost
Roles & Responsibilities	Process Map Procedures	Qualitative Supplier Team "Customer"
Functions & Tasks	Tools & Materials	Opportunities & Projects
Team Announcements	Metrics & Measures	Achievements Awards Acomplishments

FIGURE 11.2 The contents of the performance boards.

would receive a plant tour culminating in the assembly area. The tour would begin at receiving and end in the pack & ship area. Each of the 8 teams on the tour would provide the customer with an update of their team, their process and their performance. Most often, the visiting customer would witness the assembly of products they ordered during their walk-thru tour. What started out as one-customer-a-week rate, eventually became a-customer-a-day tour. We often heard the customers say, "Your competitors don't do this." These activities built customer loyalty, increased purpose and passion in the workforce, and emphasized on time, quality excellence while promoting fiduciary responsibility (i.e., since there teams were self-directed, they had to manage their own budget.) And, as I covered in Chapter 2, over a five year period, performance increased (Figure 11.3):

What was initially a loss, became a turnaround profit of $7,000,000.00. A more interesting fact, that's not apparent on the plant's performance board, is that this initiative drove a division revenue of $2.7 Billion which included software and services, as well revenue for two other divisions: the General Purpose Systems Division (Products, Software and Services) and the Customer Service and Maintenance Division.

Actual Goal Measure	1990 Baseline	1994 Outcome
Delivery Performance (Just-In-Time)	60.2 %	92%
Overall (Out of Box) System Quality	87.0%	99.2%
Customer Satisfaction	72.1%	85.6%
Associate Satisfaction	52.0%	65.5%
Capacity	72,000	98,000
Product Revenue	$220MM	$262MM
Profit (after all expenses and taxes)	-$2MM	$5MM

FIGURE 11.3 The five-year outcome of organization performance.

BEYOND PERFORMANCE EXCELLENCE

In August 2005, the Anixter warehouse in Memphis Tennessee was in the midst of a Lean transformation effort, and at the same time, preparing for a Lucent Technologies onsite assessment of their operation. That location provided the electronic components for the Lucent cell tower construction project taking place across the US. Of all the Anixter warehouses dedicated to the Lucent project, Memphis performed better than the rest. However, during Lucent's initial assessment months earlier, the Memphis location did not fare well. Anixter decided to put a program in place to address the issues at hand, increase performance and successfully pass the assessment planned for September.

After a few months of effort, Anixter Memphis increased their speed, quality and even lowered their costs. They put in place a six-process Value Stream System that managed the flow of Lucent material (i.e. Receiving → Put Away → Inventory → Pick → Pack → Ship). Using process management techniques, they increased their ability to address orders in an on-time high quality manner, while decreasing their costs by eliminating defects and errors. Using a checklist from the initial Lucent assessment, they were able to verify that they were performing above mean and expectations.

During the week of August 22, 2005, everything began as any normal Monday. During the next day, we heard that a hurricane formed in the Bahamas that had no effect and impact on Memphis for the time being. A few days later, on Friday (August 26, 2005), we were informed that a category five hurricane was heading through the Gulf of Mexico toward the Louisianna, Mississippi, Alabama coast. A strategic decision was made to load three fleets of 52' trailers with electronic and communication components focused on rebuilding the damaged communication infrastructure caused by the hurricane.

The dedicated employees of Anixter responded to the call and spent their weekend preparing the three fleets for deployment. After hurricane Katrina hit that Monday morning, August 29, 2005, the trucks headed to Louisiana, Mississippi and Alabama. In Alabama and Mississippi, the trucks were welcomed in and communication was restored a day or so later. The fleet heading into Louisianna was held up by the state police by the order of the Governor. After a couple of days, they were permitted to enter the state and deliver their material and a temporary, usable level of communication was restored a few days after Katrina landed.

The efforts of the people of Anixter Memphis impacted many lives along the Gulf coast. They completed their "normal" work for the week and spent their weekend serving the people of other states, just to return to work on Monday and go about their regular business. When performance excellence is established, the organization contains pride-in-work, flexibility and motivation to go "beyond the call of duty" when necessary, while maintain their excellent performance (Note: In September Lucent executed another assessment and Anixter Memphis scored the highest result Lucent had ever received on the Lucent Operational Assessment.).

AFFORDABILITY SCORECARD

Affordability has adopted the Kaplin and Norton "Balanced Scorecard" (i.e., Customer, Financial, Internal Business Process, Learning and Growth all centered by Vision and Strategy) basic framework at the Enterprise level for monitoring the performance of any Affordability organization. The Affordability Scorecard differs from the Balanced Scorecard in ways that convey Enterprise level performance using the dynamics of Affordability. Like the Balanced Scorecard, the Affordability Scorecard is centered and grounded by Vision and Strategy. However, when measuring Value, Value Stream and Value Delivery, the performance measures of the Customer Metrics, Business Metrics, Process Metrics and People Metrics are the key critical performance indicators of the health and status of the Enterprise. Details of each dimension are as follows:

- Customer: Satisfaction, Growth/Decline, Requirements, Profile, Markets
- Business: Revenue, Expense, Profit/Profitability, Growth/Decline
- Process: Time, Quality, Cost
- People: Satisfaction (including Suppliers and Partners), Training/Learning, Growth

At the operational level, performance of the systems and system processes with each location and every site should post and communicate the results of their scorecards. Many organizations have begun holding 15-minute huddles at the beginning of each work day to review current status, exceptions that have occurred, observed opportunities for improvement, active improvement projects, project completions and employee recognition.

More recently, organizations are starting their morning or shift at the tactical, team and individual levels with 15-minute huddles. The purpose is to be sure everyone is updated on team status and team activities. Often, team huddles are followed by management huddles (30 minutes later), which are then followed by director level huddles (30 minutes after the manager huddles). By mid-morning, or mid-shift, the entire organization is aware of the current state of the organization.

"Scorecarding" or "scoreboarding" and huddling informs every one of the current performance status and what's going on in the institution. It also serves to communicate to everyone the activities and events going on in the location. I'll repeat the old adage one more time: "You can't tell if you're winning or losing if you don't know the score." I'll add, "Everyone on the team should know the 'play' and their role and responsibility."

YOUR PATHWAY AND YOUR JOURNEY

In newly formed organizations, measuring and monitoring performance is not commonly a priority. In the beginning, survival mode is often the modus operandi. Revenue, expenses and profitability is frequently the most important priority. "Cash is King!" is the battle cry.

Soon after the beginning, and definitely when success is occurring, attention to performance should be in the forefront of your Affordability Thinking. Start with the performance of Value Delivery and Customer Requirements (Note: Assuming that financial flow and money factors are being attended to.). Value and Value Delivery should lead to the Value Stream where flow should be measured in terms of time, quality and cost. Structure performance measurement so that the Value Stream is the primary realm and the strategic, operational, tactical and team/individual layers are linked and aligned. Value Stream Support and Leadership would be next on the "docket" for architecting your performance system.

If your organization is mature, assess and evaluate how you're measuring success. Adjust and improve as necessary. Use the Affordability Scorecard to establish a linkage of performance goals and objectives throughout the organization. Measure and monitor in accordance with a reasonable rhythm (i.e., Team Performance—Daily, Process Performance—Daily, System Measurement—Weekly or Monthly, Customer Measurement—Monthly or Quarterly, Business Performance—Weekly or Monthly, Corporate Performance Consolidation—Quarterly and Yearly).

Lastly, create an environment that compares your performance to the competition (i.e., Customers, Products, Services, Processes, Workforce, Market, Competitors, etc.). Once in place, a performance excellence system supports and encourages improvement and sustainment. Be sure to use performance to recognize when performance is good. Acknowledge and award the people for achievements that are at or above the targets set. If performance falls short of expectation, adjust and modify activities according to the gaps between objectives and current state measures. In other words, "celebrate the wins, and learn from any losses" in performance.

Along the journey, performance should be an everlasting activity measured, monitored and managed by leadership to direct, align and motivate the workforce toward success, profitability and prosperity.

12 Improvement

Businesses that grow by development and improvement do not die.

– Henry Ford

IMPROVEMENT FOR AFFORDABILITY

Once some degree of Affordability has been established, continual or continuous improvement should become a regular discipline. This entails searching for opportunities for improvement that includes errors, defects, gaps and goals, and when addressed, increases performance. The primary Affordability Improvement area of concentration is Value. Since the Delivery of Value to the Customer is the fundamental AIM and purpose of Affordability, Value is the key for Customer satisfaction, loyalty, demand, growth and success.

Value can be improved by exceeding requirements, meeting unmet requirements, providing unrealized customer benefits and even reducing price. Related to any requirements input for improvement considerations, QFD, VOC and HOQ can be used to study and identify requirements not met, or unrealized requirements in the marketplace. An increase in Value can also be realized by implementing a new technology or instituting a new requirement that is beneficial but not yet present in the market. Finally, perception of greater value can occur when a price is lowered. Greater Value is fulfilled when the Customer's perception and/or understanding of the Unique Value Proposition (UVP) increases (Note: Value is Customer recognition of requirements met/exceeded, and Value perception fulfilled.).

In relation to Value, Customer improvement is possible within Affordability. Customer improvement occurs through Customer inclusion, exposure, awareness, training, knowledge, appreciation and recognition. When Customers are included, exposed and aware of the Value Systems and Value Delivery, their trust, loyalty, and the perception of reliability and dependability increase. Training and transferring knowledge of the organization and the UVP (including product and/or service details) benefits the business by exposing the Customer to the innerworkings for building up the relationship and increasing Customer awareness and opinion. When the Customer increases in positive perception improvement is realized.

The Customer/Value relationship is key to improvement and ongoing success. I learned this from Patrick Mene of Ritz Carlton in the early 1990's (Note: Ritz Carlton won the Malcolm Baldrige Award in 1992 and 1999). He coined the phrase, "We are ladies and gentlemen serving ladies and gentlemen." Ritz Carlton is definitely not a cheap hotel chain, and they do provide great value at a fair price for the customers they serve. My perception of Ritz Carlton today, is the same as it was when I first got to know them in the 1990's.

DOI: 10.1201/9781003772934-12

Of course, Cost also qualifies as a prime improvement target. To aim at Cost, both price and expense are primary targets. Price can be lowered arbitrarily but is only considered Affordable if price is lowered while margin stays the same or increases. This is where expenses and spending play a role. If expenses are reduced a greater amount than price is reduced, then the margin is increased (Note: The F/A-18 example I mentioned earlier, lowered expenses by more than $5M per jet that permitted Boeing/Northrop Grumman to lower the base price by $5M.). A focus on expense reduction does increase Affordability, and if the price remains the same, the margin is even greater. Increase in Value and the UVP, when it's appropriately conveyed to the Customer, can logically warrant an increase in price; however, it should be tested with the customer base. The two primary dimensions of Cost (i.e., Price and Expense) can be improved so that the profitability, prosperity and success can increase.

PURPOSE IMPROVEMENT

If your Value and Value Delivery is sound and the results are prosperous and successful, it's not likely that your "Purpose" needs improvement or change. However, there may be times, as I personally experienced with 9/11, that I realized my Purpose needed adjustment. When I started my company in 1995 I established the corporation under the name "People Success, Inc." The core purpose centered on People and the consulting services I offered focused on improvement of people, process, performance and success. When 9/11 occurred, and the demand for my consulting services dramatically diminished (Note: My customer project commitments went to zero almost overnight), I had to journey down another path (i.e., The Somethin's Brewin Coffee House), I was eventually able to emerge back into the consulting business when customers again started to request my services. As I emerged, I re-named my company due to some contemplation, deep thought and self-reflection, regarding new challenges and new goals that began to appear. By taking in feedback from past and present customers and mentors, I cogitated on near term possibilities and my future state. Considering my strengths, obstacles that occurred, values and principles (Note: That didn't change), and past business experiences, I named the new version of my organization "Performance Excellence Associates, Inc." I must add that much of the content, concepts and concentration of the business didn't change, but the new approach to my journey brought me to the Theory of Affordability.

PEOPLE IMPROVEMENT

The word "people" typically refers to the plural case of individuals or persons. However, the basics of the improvement of people is based on the improvement of all individuals. This involves the training and development of each and every person. Training is the transfer of knowledge and some capability, while development encompasses the advancement of the person in terms of the physical, intellectual, emotional, spiritual and cultural dimensions and aspects of every human.

For training, I have 50 Topics for knowledge transfer that I've compiled over the last 30 years (Note: You may have some additional subjects to add): use of time, efficiency and organization, communication, researching, writing, interpersonal behavior, integrity, initiative, ambition, cooperation and collaboration, teamwork, ability and capability, workplace competency, adaptability, flexibility, collaboration, confidence, effectiveness, service (customer, supplier, people), analytical and critical thinking, problem solving, decision-making, negotiation, delegation, leadership, organization and structure, goal setting, alignment, planning, targeting and aiming, measurement, performance improvement, executing, conflict resolution, listening, speaking, patience, proactiveness, advocacy, perseverance, stress, responsibility, accountability, feedback (receiving, giving), mentoring, coaching, teaching, improving (people, process, performance), innovation and invention, learning and continuous learning, Affordability.

For development, I prefer experiential learning (which I often include in the training programs I offer), that is designed to engage the body, mind, heart and inner spirit of the individual, individuals and teams. I like to use games, simulations, problem solving challenges, purposeful projects and topic focused workshops. For topic areas of Lean, I've used logo team building, 3P events, 5S activities, Green Belt Projects and Black Belt Projects. For Six Sigma, I've used dice rolling, catapult activities, DMAIC simulation, as well as Six Sigma Green Belt and Black Belt Projects.

By far, the best method of training and developing people is one-on-one. Training transfers knowledge and development involves experiential learning, and both can be accomplished in a one-on-one setting. Beyond training and development, lies coaching and mentoring. Coaching is the development of skills one-on-one. Mentoring involves the alignment of the individual's purpose one-on-one. Once a common practice in business in many institutions during the 20th century, during 'onboarding' of an employee, a coach and mentor was assigned to every new associate. The coach/mentor served as a guide, confidant, teacher, coach and mentor. Recently, I've seen a slight resurgence of this practice, but I believe it can be more widely used. I recommend this as a standard practice in organizations pursuing Affordability.

Although the topic of management has had a great deal of popularity over the past 50 years, the topic of leadership has been neglected in higher education, corporations and our society. This is the one topic of training and development I emphasize with every organization I serve.

PROCESS IMPROVEMENT TOOLS AND TOOLBOXES

Process improvement is a key critical component of Affordability. It involves Suppliers, the Value Stream, Value Delivery and each and every Value Stream Support function and Leadership Support area. Financial processes, technical support processes, human resource support processes, supplier support processes, customer support processes, marketing and sales processes and all other processes of the enterprise all qualify for improvement just as much as the Value Stream processes of creation, production and delivery. Although there are many toolboxes and tools available for improvement initiatives, there are a few that I have heavily relied on since the 1990's. From my original experience with Just-In-Time beginning in the late 1980's, the tools of Lean are my often "go to" toolbox when beginning a process improvement project. After the process has been documented, stabilized, standardized, and is predictable, I use the Six Sigma tools in the 6σ toolbox. Throughout many projects, I have used the quality tools from the TQM toolbox I learned about from Bell Labs and the ASQ.

When beginning many process improvement projects, after observing, researching and assessing the current state, I most often found myself using the Lean tools of the Lean toolbox to create quick wins for project momentum focused on eliminating waste and reducing variation. The process improvement strategy for Lean is simple and straight forward. Its and four step approach that's easy to remember and easy to use. It flows as follows:

1. Current State → 2. Improvement Opportunities → 3. Future State →
4. Implement Improvements

With a process improvement team, first establish the current state of the process. This includes documenting and mapping the processes (i.e., Value Stream Mapping and Process Mapping), understanding the variables and factors that function in the process, measuring and monitoring the performance metrics of the process, and comprehending the entire people → process → performance relationships.

The next step is to identify the process improvement opportunities to eliminate waste, reduce variation and improve performance. Since I always recommend a team for process improvement efforts, I put the team to work on this task. Taking what they know and have discovered from step 1, I bring them to the Gembe (Note: Japanese term for the place where it happens) and have them use the artifacts they created and their knowledge to identify the opportunities for improvement.

The third step is to have the team render a "future state" of the process of what the process should look like, how it should perform with what needs to be done to improve performance. Improvement opportunities should be documented with solutions and implementation plans should be created.

Finally, the implementation plans created should be executed and the improvement implementations should be realized. At this point, process performance should be monitored and measured for improvement opportunities and targets. At this point, the future state, now becomes the current state (Note: Then there's no need to render a current state since the future state has now become the current state), and the team can move on to step 2 to identify additional improvement opportunities, then step 3, then step 4, then step 1 to 2, and the cycle of continual improvement continues.

At the enterprise level, the overall awareness throughout the organization of Lean, Lean Thinking and Lean Principles should be established. Lean is comprised of five tenets (Figure 12.1).

From this perspective, Lean Practices for Affordability can be instituted, and the Lean toolbox and tools can be established throughout the Value Stream, Value Stream Supports functions and Leadership Support. After Lean is implemented, Six Sigma requires that systems and processes be documented, stable, standard and predictable.

The Six Sigma toolbox contains some of the Lean tools from the Lean toolbox, with additional statistical tools and Six Sigma specific tools. The standard problem-solving process for process improvement is most often referred to as D.M.A.I.C. representing Define, Measure, Analyze, Improve and Control (Note: The D.M.A.I.C. version I use comes from the IISE standard D.M.A.I.I.C., which includes a critical sub step of Implement that ensures that a plan, execution and verification happen.) (Figure 12.2). Below is the DMAIIC Toolbox with many of the tools affiliated with

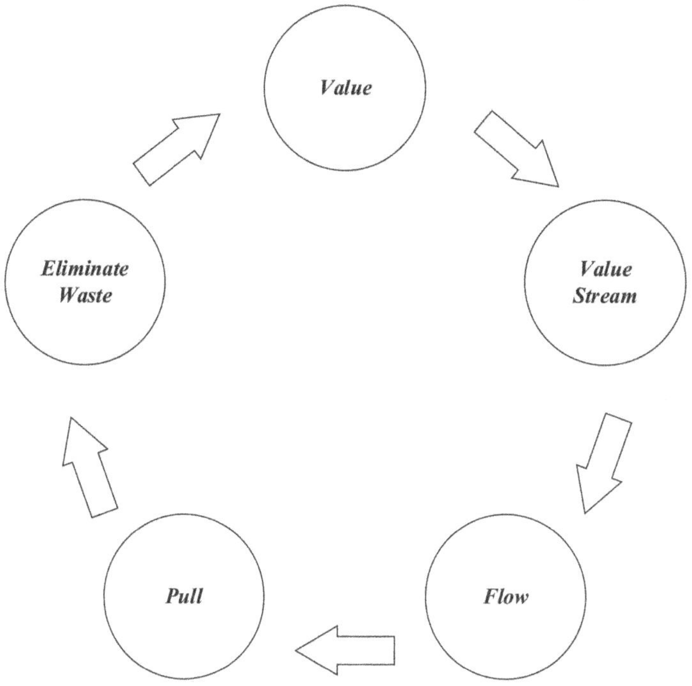

FIGURE 12.1 The five tenets of lean.

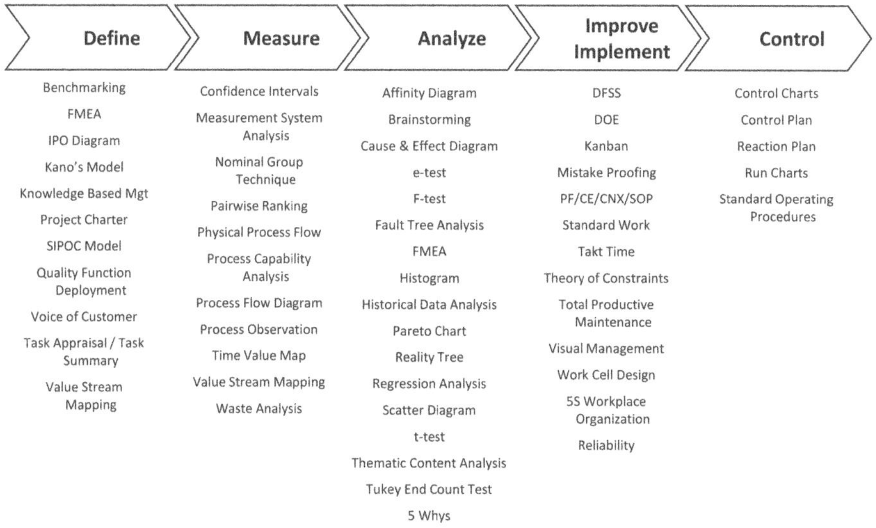

FIGURE 12.2 The D.M.A.I.I.C. Problem solving process with affiliated tools.

each phase listed (Note: Some tools are listed more than once because they may occur in a couple of the phases):

Using Six Sigma (AKA 6σ) for process improvement, a more detailed statistical approach is orchestrated. There was one organization I served began with 6σ initiatives for system and process improvement, and after two years abandoned that approach to re-start with Lean. The Lean initiative lasted 12 years until the leadership team determined it was "6σ time" and they resumed the 6σ effort. They are now one of the best Healthcare Systems in Georgia.

The third process improvement toolbox I'd like to feature involves both basic and advanced quality tools. As the Director of Quality for NCR's Retail Systems Division in the early 1990's I learned about 14 powerful quality tools from my relationship with the ASQ. I was able to experience these tools while working at NCR and use the toolbox afterward for client process improvement efforts.

The "7 Basic Quality Tools" for Total Quality Management (Reference: ASQ, from its origin, the 7th tool has evolved.) is as follows:

1. Cause and Effect Diagram (AKA: Fishbone Diagram or Ishikawa Diagram): Method for pursuing root causes.
2. Check Sheet (AKA Checklist): A structured form for identifying, collecting, verifying and analyzing data.
3. Control Chart: Statistical tool for displaying and monitoring process stability and performance special cause variation.
4. Histogram: a visual, graphical representation of data distribution.
5. Pareto Chart: A combination chart using bars to illustrate the most significant and least significant elements in a data set and line to show the cumulative quantity of the data.
6. Scatter Diagram (AKA Scatter Plot or Scatter Graph): Is a graph that represents the relationship between two variables.
7. Stratification (Refers to breaking down data into categories to more easily make sense of them) or Flowchart (A visual diagram illustrating the order and flow of process using specific shapes for process functions) or Run Chart (Line chart that displays data over time for identifying trends, shifts and randomness).

The "7 Advanced Quality Tools" often used and utilized in conjunction with DMAIIC efforts and process improvement projects.

1. Affinity Diagram: Organizes large amounts of data, ideas or opinions into meaningful groups for identifying patterns and relationships, and analysis of data groupings.
2. Interrelationship Diagram (AKA Relations Diagram): Visualizes cause-and-effect relationships among multiple factors for identifying root causes in system interactions.
3. Tree Diagram: Breaks down broad goals into detailed, actionable steps for planning project management and process detailing.

4. Matrix Diagram: Shows relationships between two or more sets of data, and evaluates the strength of those relationships for analysis and prioritization.
5. Prioritization Matrix: Ranks options based on weighted criteria and facilitates decision-making when multiple alternatives exist.
6. Activity Network Diagram (AKA Critical Path Method): Maps activities, dependencies and timelines in an effort to identify the critical path of a project.
7. Process Decision Program Chart (PDPC): Used for anticipating potential problems by breaking down processes into steps for risk identification for developing contingencies.

PERFORMANCE IMPROVEMENT

Performance improvement is about using the measures of system and process output to identify opportunities, plan for change and enhancement, and execute transformation initiatives. There are three critical areas of Affordability that can be considered focal points for overarching performance improvement (i.e., Systems, Structure, Strategy). Below are some brief descriptions of where and how performance improvement can be concentrated:

- Systems: Supplier, Value Stream, Customer areas are all enterprise-level Affordability systems areas. Value support systems and leadership systems are also viable targets for performance improvement. To improve these systems, the people and processes involved in each of these systems should be assessed for an understanding of the opportunities for improvement. Whether a Lean approach is taken (i.e., Current State → Identification of Improvement Opportunities → Future State → Implement Improvements), or Six Sigma methods are employed (i.e., DMAIIC), or the TQM Toolbox and Tools are exercised, or a combination of any of the three are utilized, performance improvement should be a continuous objective for the enterprise.
- Structure: As it applies to the Value Stream construct, or the organization's composition, must align with Affordability. Alignment of resources is a top action of leadership and is the responsibility and accountability of the leadership team. The enterprise structure both physically and functionally must be directed, aligned and motivated to performance deliver value to the customer.
- Strategy: Also, a leadership responsibility. The leadership team is accountable under the Affordability design to set strategy, execute strategy, monitor the overall current performance, and strategize improvements. The strategic plan should serve to communicate and direct execution.
- Strategy Tools: Some tools that can be used for creating strategy include:
  - SWOT Analysis: An analysis approach using four dynamics (i.e., Strengths, Weaknesses, Opportunities, Threats). After analyzation, plans are put in place to address and leverage each of the four critical areas.
  - GAP Analysis: A systematics approach and process to compare an organization's current performance with its desired future state for the identification of deficiencies and create actionable plans for improvement.

- Porter's Five Forces: A strategic framework for analyzing competitive forces impacting performance and profitability: 1. Competition, 2 New Competition or "Market Entrants," 3. Suppliers and their 'Bargaining Power', 4. Buyers and their 'Bargaining Power', 5. Threats of Substitutes. Analysis and mitigation of the five forces strategically fit in the enterprise strategic level of leadership thinking.
- Balanced Scorecard: As adopted from Kaplin and Norton's "Balanced Scorecard" (i.e., Centered on Vision and Strategy—Customer, Financial, Business Processes, Growth and Learning). Affordability: Customer, Business, Process, People.
- S.M.A.R.T.: A goal setting tool using framework based on Specific, Measurable, Achievable, Relevant, Timely factors.
- PESTEL: A tool for conducting an "environmental scan" using several viewpoints: Political, Economic, Social, Technical, Environmental, Legal.

Each of the tools mentioned provides a variety of perspectives regarding performance improvement. From the 5P model, the primary input for performance comes from people and processes. For performance to improve, people and process improvement would likely be involved. Improvement of performance leads to improvement of profitability and prosperity. Leaders must be ever-vigilant on performance, process and people to realize improvement of profitability and prosperity.

POTENTIAL FOCAL POINTS FOR IMPROVEMENT

Continual or continuous improvement is achieved through constant focus and attention on those opportunities that yield performance enhancement and advancement. Beyond the tools, toolboxes and techniques already mentioned, there are specific potential areas that can be observed, analyzed and understood to identify actionable improvement targets. For the full effect of improvement to be realized, multiple general 'causes' contain the primary 'roots' that may serve as foci of efforts and activities that lead to improvement. A unique use of the cause-and-effect or fishbone diagram can be used to identify and focus efforts to create and implement solutions (Figure 12.3).

Manpower: People, already mentioned earlier, is a good starting point for identifying improvement opportunities. Beyond training and skills, cultural aspects of associates in the workplace (e.g., values, principles, interrelationships and teamwork) are success factors in any enterprise. Customer service is an attribute beneficial to every Value provider. Identifying areas of manpower improvement is the first step in understanding the gaps and shortfalls of workers. Remember: Current State → Improvement Opportunities → Future State → Improve.

Methods: Work methods of processes, procedures and tasks are an ongoing opportunity for improvement. Continuous improvement (C.I.) of systems and processes should be in the "muscle memory" of the organization. As a part of every workday, individuals and teams should have C.I. inculcated in their daily rhythm and regimen.

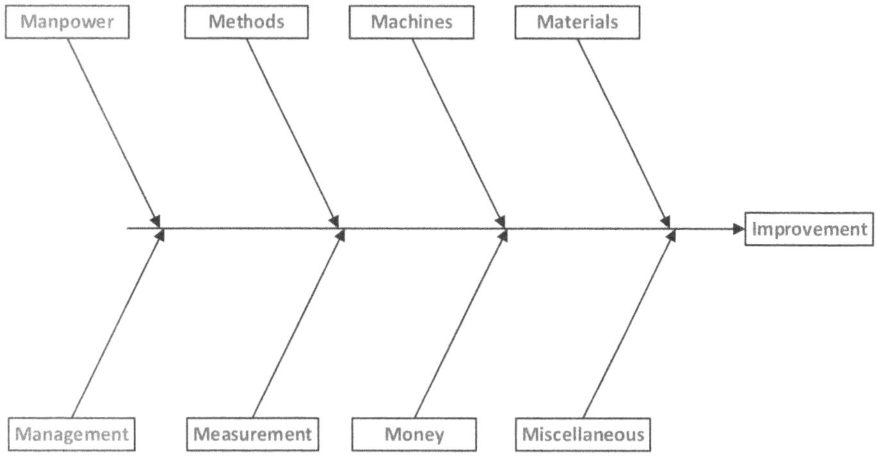

FIGURE 12.3 An Ishikawa diagram illustrating primary root causes for improvement.

Materials: Beyond product and service materials, materials that support the worker in executing their jobs qualify as an opportunity for improvement. The availability and accessibility of materials increase efficiency, decrease time wasted with improved performance. The flow of any, and all types of material is often overlooked when measuring performance. This includes the right amount, not too much, not too little.

Machines: Machines, technology and tools are critical factors contributing to success. Researching and studying the proper devices and instruments required and necessary identifies potential areas for improvement. Understanding the employee competency levels for these mechanisms and apparatuses also serves in the effort of discovery for improvement.

Management: The proper management of operational aspects and systems leads to profitability and prosperity. Those who manage the "things" of the enterprise must be effectively prepared to supervise and oversee business operations. Management can be a hinderance just as much as an asset. Improvement of management directly affects people, process and performance.

Measurement: As an integral part of Affordability, measurement of the enterprise, the systems, the processes and the performance is critical. Basic measures of time, quality, cost and satisfaction are a requisite. An improvement opportunity exists on the determination of the proper metrics, goals, targets and measuring methods. Qualifying the merit of each measure and metric is another potential matter to pursue.

Money: The use of money, the waste of money, the impact of money and all factors of money is always a topic for research and discovery for improvement. Money is neither a problem nor a solution, but it is a tool the leverage, measure and utilize.

Miscellaneous: Topics that may qualify in this category include: Value, Customers, Suppliers, Competitors, Environment, Location and those "other categories" that are suspect for the improvement opportunities list.

For any improvement endeavor, the overarching goal for Affordability is to eliminate waste, increase efficiency and effectiveness, improve performance and optimize the Value, Value Stream and Systems that generate profitability and prosperity.

HOW ENGINEERING HELPS

Around 20 years ago, I formed a partnership with the Institute of Industrial and Systems Engineering (AKA IISE) for the purpose of deploying programs that educate, prepare and certify individuals in the areas of Lean, Six Sigma and other engineering approaches to improvement (e.g., Project Management, Professional Engineer, Quality Tools). During that time, we served over two dozen universities and more than fifty organizations. Most of the effort focused on improvement of the individuals participating, and ultimately, improvement of the organizations they represent.

The first pilot program was tested and validated at Georgia Tech with a small group of Industrial and Systems Engineering Seniors. It entailed three days of knowledge transfer, with several simulations and a final exam to earn a Lean Green Belt Certification (Note: IISE programs are accredited by ABET and IACET). This program was subsequently rolled out to more than 20 universities across North America certifying literally 1,000's of college students (i.e., other than industrial engineers, mechanical engineers, electrical engineers, nuclear engineers, civil engineers, chemical engineers, petroleum engineers, mining engineers, biochemical engineers and business students).

After the Lean Green Belt (AKA LGB) Program was validated and verified, it was then utilized in a variety of businesses and industries: Manufacturing and Assembly, Insurance, Healthcare, Architecture, Telecommunications, Banking, Distribution/Logistics, Warehousing, Package and Mail Delivery, Pharmaceuticals, Oil, Energy, Construction, Food Production, Food Service, Government. Employees from various companies attended IISE workshops, and a multitude of companies contracted with the IISE to deliver private workshops. Often, associates attending the sessions were required to complete an improvement project as part of the certification criteria.

Shortly after the Lean Green Belt Program was in place, a Lean Black Belt (AKA LBB) Program began. In parallel, Six Sigma Green Belt (SSGB) and Black Belt (SSBB) Programs were created and eventually Lean Six Sigma Green Belt (LSSGB) and Black Belt (LSSBB) Programs were finalized and instituted. This portfolio of improvement programs based on an engineering body of knowledge generated a great deal of improvement from 2006–2026, by both the certified students who graduated, and the professionals who were certified.

One specific initiative involved the Georgia Hospitals Association (AKA GHA) who leveraged a training grant to hold Lean Six Sigma Greenbelt (AKA LSSGB) and Lean Six Sigma Blackbelt (AKA LSSBB) certification sessions. There are 187 hospitals in the State of Georgia with approximately 53 community health and hospital systems. Mostly, all the healthcare sites in Georgia participated in certifying 1000's of LSSGB's and 74 LSSBB's. The certification projects of the 74 LSSBBs returned a cost reduction total of $57,409,457.00 (Avg Savings per Project was $775,803.47), as well as performance increases in the areas of time, speed, accessibility, quality and especially patient satisfaction.

As a result of these programs and projects, it has been proven that improvement efforts can increase performance in terms of time, quality, cost, Customer Satisfaction, Employee Satisfaction and Supplier Satisfaction while eliminating waste and reducing variation while increasing profitability.

IMPROVEMENT PROJECTS PAY-OFF IN TERMS OF VALUE AND WORTH

In 2011, as a result of Medivators' acquisition of Byrne Medical, the newly formed organization wanted to institute Lean Thinking, Lean Practices and Lean Principles in their engineering and manufacturing operations. The plan was to train all executives (i.e., Lean and Leadership), the team leads and managers in manufacturing and the employees in the assembly areas in LGB requiring a completed improvement project for certification (Note: The project requirement was a specific prerequisite for this organization's certification.).

After the team leads and managers attended a three-day LGB Program and passed the exam, they were obligated to submit a project proposal and project charter, and once approved, the project had to be completed and accepted by the executive staff before a certification was granted. Since every system and process in the Value Stream (i.e., creation, production, delivery) was involved in LGB Projects, the entire Enterprise from Suppliers to Customers was under some activity for improvement. After a couple of years, the ROI was paying for itself. The Value increased, the Customer Satisfaction increased, the Supplier Satisfaction increased and the Employee Satisfaction increased. For the next several years, the success, profitability, prosperity, value and worth of Medivators increased dramatically.

NOTE: Since the Value of Medivators increased so substantially since the Medivators acquisition of Byrne Medical in 2011, Cantel Medical was so attracted by the company's increased worth that they acquired Medivators in 2021 as part of the pending merger with Steris during the latter part of that year.

Please keep in mind that the means, methods, tools and toolboxes presented in this chapter only represent a portion of all the approaches that can be taken to achieve Affordability Improvement results. I highly recommend that you take a deep dive into all aspects of the improvement marketplace.

YOUR PATHWAY AND YOUR JOURNEY

Improvement initiatives should begin when Affordability is being established. Implementing Affordability is the initial effort to improve the organization. Attention to improvement helps accomplish system functionality to achieve a better tempo and performance rate within the Value Stream and Value Stream Support processes. Improvement is all about the endeavor to "make better."

Before delving in too deep or "jumping the gun," take some time to observe, research, study and seek to understand the opportunities for improvement. Create the actual current state and identify opportunities for improvement and define the targeted areas that need to be improved. Be sure to prioritize your actions and pursue quick wins in the beginning to establish momentum and motivation. Instead of making one improvement at a time, create solution teams to address multiple improvements simultaneously. Remember, teams of 5–8 are the optimum team sizes for problem solving efforts.

On a broader basis, create a strategy for improvement of all of the organization. As a leader, be involved and engaged in the problem-solving efforts, even to the extent of participating on a problem-solving project. Demonstrating commitment always influences attitudes of others in terms of participation.

Always seek development and improvement of the Value you deliver, your Value Stream and the Enterprises Performance.

13 Sustainment

When Affordability is understood, pursued, reached, accomplished and established, Sustainment is a necessary vigil for ongoing improvement and continued existence.

– Paul Walter Odomirok

Success and sustainability go hand in hand when done right.

– Unknown

Sustainability is not a goal to be reached but a way of thinking.

– Unknown

SUSTAINMENT OF AFFORDABILITY

Sustainment of Affordability is a critical dimension of the Theory of Affordability and Application of Affordability. The definition of sustainment, and its variants, according to Webster are:

- Sustain: cause to continue or be prolonged for an extended period or without interruption.
- Sustaining: providing physical or mental strength or support.
- Sustained: continuing for an extended period or without interruption.
- Sustainability: the ability to be maintained at a certain rate or level.
- Sustainment: the support or maintenance of someone or something, especially military equipment or personnel (Note: And for Affordability, Enterprises and Governments).

Once you've reached a level of performance excellence (in accordance with your goals) and can verify Affordability using the Success Formula and Affordability Factors as validation criteria, an Affordability assessment can be performed to confirm your status and begin to implement and deploy sustainability practices.

I have a simple philosophy to follow: "Maintain and continue what's going well and discover what can be done to make things even better." Improve Value, strategy, products, services, systems, processes, structure, competitiveness and 'footprint' upon increased demand and opportunities. Continue to research and seek different aspects of the marketplace that might accept variations of your Value, as well as other potential markets that may embrace new versions (e.g., the technology for GPS products invented for locating positions were acclimated for automobiles, walkers, runners and bicyclists, or the use of Lean Thinking from Toyota in healthcare, military logistics and maintenance, and even with the

DOI: 10.1201/9781003772934-13

manufacturing and assembly of products other than automobiles). Investigate, experiment, pilot and validate other forms of the Value you offer for additional markets. In parallel, examine other establishments and institutions for tips and pointers regarding how they've maintained and sustained their companies and establishments.

After I wrote the book "Affordability" in 2016 and published it in 2017, and moved to St Augustine on November 1, 2018, then subsequently published the book "Healthcare Affordability in 2019, I began to investigate, identify and research local businesses and enterprises that demonstrate Affordability and the Sustainment of Affordability. I was surprised to discover that there are numerous examples that have existed in St Augustine's from its founding in 1565 until today. Below are just some of the many samples of my discoveries.

AFFORDABILITY SUSTAINMENT: THE OLDEST AND NEWEST EXAMPLES I'VE EXPERIENCED RECENTLY

St Augustine, the oldest continually occupied and operated settlement in the United States, was founded more than 460 years ago in 1565. I moved to St Augustine in November 2018, where my in-laws lived since 1981. When I arrived, it was very Affordable, however its level of Affordability has changed since COVID struck the US. Since I was first exposed to the city, I became quite familiar with St Augustine (AKA: "STA") over the past 45 years. I've been able to study its culture, the socio-economics and business environment as it evolved and changed. To understand its sustainability, its worthwhile understanding its history.

From its founding in 1565, St. Augustine has earned the moniker as "the oldest continuously occupied settlement of European and African origin in the United States" (Note: 42 years before the English colonized Jamestown and 55 years before the Pilgrims landed at Plymouth Rock). As adopted from input of the City of St Augustine, there have been 16 distinct historic periods for the community:

- 1513 to 1565: The Discovery Period of Ponce de Leon
- 1565 to 1763: First Spanish Colonial Period
- 1763 to 1784: British Colonial Period
- 1784 to 1821: Second Spanish Colonial Period
- 1821 to 1845: U.S. Territorial Period, Florida becomes a state
- 1845 to 1861: Early Statehood Period
- 1861 to 1865: U.S. Civil War
- 1865 to 1885: Post-Civil War Period
- 1885 to 1913: Flagler Era
- 1913 to 1919: World War I Era
- 1920 to 1926: Boom Time
- 1926 to 1941: Depression Era (Florida)
- 1941 to 1945: World War Two
- 1945 to 1970: Post WWII
- 1970 to 2000: Changing Economics, No Growth
- 21st Century: Population Growth and Success

Some highlights:

- Traditional Architecture: The existing architectural legacy of the city's past is much younger, than the earliest wooden structures destroyed by invading British forces in 1702 (Note: Only the coquina stone fort, Castillo de San Marcos, completed in the late 17th century, survived). Vestiges of the First Spanish Colonial Period (1565 to 1764) remain today in St. Augustine in the form of the town plan originally laid out by Governor Gonzalo Méndez de Canzo in the late 16th century and in the narrow streets and balconied houses that are identified with the architecture introduced by settlers from Spain. Throughout the modern city and within its Historic Colonial District, there remain 36 buildings of colonial origin and another forty that are reconstructed models of colonial buildings. St. Augustine can boast that it contains the only urban nucleus in the United States whose street pattern and architectural ambiance reflect Spanish origins.
- Discovery: Historians credit Juan Ponce de Leon, the first governor of the Island of Puerto Rico, with the discovery of Florida in 1513. While on an exploratory trip in search of the fabled Bimini he sighted the eastern coast of Florida on Easter Sunday, which fell on March 27 that year. Ponce de Leon claimed Florida for the Spanish Crown and named it Florida after the Easter season, known in Spanish as Pascua Florida. This newly claimed territory extended north and west to encompass most of the known lands of the North American continent that had not been claimed by the Spanish in New Spain (Mexico and the Southwest).
- Settlement: In the following half century, the government of Spain launched no less than six expeditions attempting to settle Florida; all failed. In 1564 French Huguenots (Protestants) succeeded in establishing a fort and colony near the mouth of the St. Johns River at what is today Jacksonville. This settlement posed a threat to the Spanish fleets that sailed the Gulf Stream beside the east coast of Florida, carrying treasure from Central and South America to Spain. As Don Pedro Menéndez de Avilés was assembling a fleet for an expedition to Florida, the French intrusion upon lands claimed by Spain was discovered. King Philip II instructed Menéndez, Spain's most capable admiral, to remove the French menace to Spain's interests.
- Naming St. Augustine: On September 8, 1565, with much pomp and circumstance and 600 voyagers cheering, Menéndez set foot on the shores of Florida. In honor of the saint whose feast day fell on the day he first sighted land, Menéndez named the colonial settlement St. Augustine. Menéndez quickly and diligently carried out his king's instructions. With brilliant military maneuvering and good fortune, he removed the French garrison and proceeded to consolidate Spain's authority on the northeast coast of Florida. St. Augustine was to serve two purposes: as a military outpost, or Presidio, for the defense of Florida, and a base for Catholic missionary settlements throughout the southeastern part of North America.
- Military Colony: Maintaining St. Augustine as a permanent military colony, however, was a mighty task. Without the courage, perseverance, and

tenacity of the early settlers, it is doubtful that the community would have survived. English pirates and corsairs pillaged and burned the town on several occasions in the next century. Clashes between the Spaniards and the British became more frequent when the English colonies were established in the Carolinas, and later, in Georgia. Consequently, the Spanish moved to strengthen their defenses, and in 1672 began construction of a permanent stone fortress. The Castillo de San Marcos was brought to completion late in the century, just in time to meet an attack by British forces from the Carolinas in 1702. Unable to take the fort after a two-month siege, the British troops burned the town and retreated.

- Underground Railroad: During the 1700's, plantation and slave owners in the English colonies resented the sanctuary that Spanish Florida afforded escaped slaves who successfully made their way to St. Augustine, which became a focal point for the first Underground Railroad. Escaped slaves were given their freedom by the Spanish Governor if they declared allegiance to the King of Spain and embraced the Catholic religion. In 1738 the first legally sanctioned free community of former slaves, Gracia Real de Santa Teresa de Mose, was established as part of the presidio's northern defenses. In 1740, an even stronger attack on St. Augustine was mounted by the Governor of the British colony of Georgia, General James Oglethorpe. He also failed to take the fort.
- Treaty of Paris: The Treaty of Paris in 1763, ending the French and Indian War, gave Florida and St. Augustine to the British, accomplishing by the stroke of a pen what pitched battles had failed to do. St. Augustine came under British rule for the first time and served as a Loyalist (pro-British) colony during the American Revolutionary War. A second Treaty of Paris (1783), which gave America's colonies north of Florida their independence, returned Florida to Spain, a reward for Spanish assistance to the Americans in their war against England. Upon their return, the Spanish in 1784 found that St. Augustine had changed. Settlers from a failed colony in New Smyrna (south of St. Augustine) had moved to St. Augustine in 1777. This group, known collectively as Minorcans, included settlers from the western Mediterranean island of Minorca. Their presence in St. Augustine forever changed the ethnic composition of the town.
- Second Spanish Period: During what is called by historians the Second Spanish Period (1784 to 1821), Spain suffered the Napoleonic invasions at home and struggled to retain its colonies in the western hemisphere. Florida no longer held its past importance to Spain. The expanding United States, however, regarded the Florida peninsula as vital to its interests. It was a matter of time before the Americans devised a way to acquire Florida. The Adams-Onîs Treaty, negotiated in 1819 and concluded in 1821, peaceably turned over the Spanish colonies of East and West Florida and, with them, St. Augustine, to the United States.
- Florida Becomes a State: In 1845 was Florida accepted into the union as a state. The Territorial Period (1821–1845) was marked by an intense war with native Indians, the so-called Second Seminole War (1835–1842). The

United States Army took over the Castillo de San Marcos and renamed it Fort Marion.

- Civil War: In 1861, the Civil War began. Florida joined the Confederacy, but Union troops loyal to the United States Government quickly occupied St. Augustine and remained in control of the city throughout the four-year long war. St. Augustine was one of the few places in the United States where Abraham Lincoln's Emancipation Proclamation, issued in 1862, actually freed any slaves. After the war, land was leased to freed slaves on what was then the west bank of Maria Sanchez Creek. Initially called Africa, the settlement later became Lincolnville and is today listed in the National Register of Historic Places, along with three other historic districts in the city.
- Vacation Town: 20 years after the end of the Civil War, St. Augustine entered a new era. Following a visit to the crumbling old Spanish town, Henry Flagler, a former partner of John D. Rockefeller in the Standard Oil Company, decided to create in St. Augustine a winter resort for wealthy Americans. He owned a railroad company that in 1886 linked St. Augustine by rail with the populous cities of the east coast. In 1887, his company began construction of two large and ornate hotels and a year later added a third that had been planned and begun by another developer. Flagler's architects changed the appearance of St. Augustine, fashioning building styles that in time came to characterize the look of cities throughout Florida. For a time, St. Augustine was the winter tourist mecca of the United States.
- Newport of the South: In the early 20th century, the very rich found other parts of Florida to which they could escape. With them fled Flagler's dream of turning St. Augustine into the "Newport of the South." St. Augustine nevertheless remained a tourist town. As Americans took to the highways in search of a vacation land, St. Augustine became a destination for automobile-borne visitors. The tourism industry came to dominate the local economy.
- Restoration: The city celebrated its 400th anniversary in 1965 and undertook in cooperation with the State of Florida a program to restore parts of the colonial city. The continuation of an effort actually begun in 1935, what became known as the "Restoration" resulted in preserving the 36 remaining buildings from the colonial era and the reconstruction of some forty additional colonial buildings that had previously disappeared, transforming the appearance of the historic central part of St. Augustine. It was in great part a tribute to such efforts that King Juan Carlos and Queen Sofia made this small city a part of their 2001 visit to the United States.
- Civil Rights Era: In 1964, St. Augustine played a role in America's civil rights struggle when Dr. Martin Luther King, Jr. led a local campaign to dramatize national efforts designed to secure Congressional approval of what became the landmark Civil Rights Act of that year. The city now contains a series of historical markers noting sites associated with the civil rights movement here. This past month, a bust of Frederick Douglas was unveiled in the downtown square by Governor DeSantis.

- St. Augustine Attracts Visitors: Some two million visitors annually make their way to St. Augustine, lured by the sense of discovering a unique historic part of America. While the venerable Castillo de San Marcos remains the traditional magnet for visitors, there are many other appealing historical sites and vistas. The City of St. Augustine maintains architectural control over the colonial city, ensuring that the inevitable change which occurs in a living urban area respects the past (Worthy of Noting: In response to the increase of shrimp harvesting in the 1940's and 1950's, the city was called "Shrimp Boat City" because of the shrimp boat production industry which lasted into the 1980's.)

Here are some local Businesses of St Augustine that demonstrate Affordability Sustainment.

- Flagler Hotels: Beginning in 1885, imagined, constructed and operated by Henry Flagler beginning more than 141 years ago. Some highlights and milestones include: one of the first poured concrete building in the world, one of the first buildings in the country wired for electricity and powered by generators installed by Thomas Edison, the largest collection of tiffany glass windows, running water in all rooms, first hotel in Florida, rail transportation to the hotels from other US locations by way of Flagler's Florida East Coast Railway. As time passed, each site evolved as the customers and the community change and transformed, and demands placed on those sites shifted in terms of the Value they could offer and deliver. The first of Henry Flagler's three great hotels, the Ponce de Leon, was adapted for use as an institution of higher learning in 1971. As Flagler College, it expanded to embrace a student body of some 1,700 by the end of the century, offering a traditional four-year arts and science degree program that serves 2,441 students at the institution today. The second of his hotels, the Alcazar, since 1948 contained the Lightner Museum and beginning 1973 also contains some of the City of St. Augustine's municipal offices. The third Flagler hotel, originally called the Casa Monica, operated as a high-end hotel until 1930, then stood vacant for 35 years before St. Johns County converted it for use a county courthouse in 1965, then in 1999, under private ownership, the building was restored to its original function and is now the only one of Flagler's three great hotels still serving the purpose of providing hospitality. Its worthy to note that each of the buildings created by Henry Flagler changed and adjusted their purpose and business charter over time in concert with the needs, requirements and demands of the St Augustine Customers and STA Community.
- The St Augustine Alligator Farm Zoological Park: Opened May 20, 1893, on St. Augustine Beach as a minor tourist attraction where visitors could go to see alligators and purchase souvenirs. Its current location was established in the 1920's by the original owner, and purchased by the current owners, the Drysdale family, in 1937. Today, David Drysdale, a family

patriarch, heads the park which contains 24 species of crocodilians, numerous reptiles, mammals and birds (there's even a wild bird rookery for native species). The cost for a family of four (i.e., two adults, two children), ranges from $92.97 (two adults and two children under 3), to $113.96 (two adults and two children 3–11), and $185.94 (two adults and two children 12 +). Although the admission is not "cheap," it is of good Value, and the experience is worth the expense. Features include: natural habitat exhibits, animal feedings, animal events, lectures and educational demonstrations (Note: They even offer zip lining above the park). Most visitors spend several hours, if not the whole day walking, observing, eating, drinking, learning, experiencing and souvenir shopping. They attract over a million visitors each month from around the world. As for Affordability and sustainability, they've been in business over 130 years and they check all of the boxes of the Success Formula and Affordability Factors.

- Price's Barber Shop: Started in 1936 by the Price's, now 90 years old, and led by Mike Price today. They service mostly men with six traditional barber chairs in an atmosphere similar to the nostalgic tonsorial parlors of old. It's a "no frills," cash only business that provides custom cuts from very talented barbers and stylists (Note: About half of the employees are men, and half of them are women). Their prices match local and corporate competitors, without the pomp and circumstance of flashy nuevo "bells and whistles." The employees earn half the price of the cut cost plus tips, and they average 10 to 20 cuts per day per barber, services 100 or more customers each day. The Value is high, the overhead is low, and the customer base is loyal. They know their Value, they know their customers, they know their competition and they know the dynamics of Affordability.
- Whetstone Chocolate: After opening a small ice cream shop on the old Saint George St in St. Augustine in 1966, Henry and Ester Whetstone founded a chocolate company in 1967 (soon to be 60 years old next year). Virginia Whetstone, their daughter, is now the CEO of the company and "Hank Sr" continues to be involved today (Note: His son, "Hank Jr," is the CEO of The Atlantic Candy Company.). Their target customer base is the local community, visitors and vacationers from out of town. Other than selling a wide variety of chocolate-based candies (Note: They even offer alligator and dolphin shaped chocolates on a stick) and fudge, they sell chocolate derivative products (i.e., syrups, powders, chocolate covered fruits, nuts and coffee beans), a small amount of other candies, offer gifts in their three strategic locations, and tasting tours in their production facility in downtown St. Augustine. Their Value is high, their prices are reasonable and they've become a mainstay in the community. Their sustainability ingredients and formula are strong.
- Kyle's Seafood: A local seafood market featuring the freshest seafood was started more than 28years ago by Mike Kyle. Recently, they were featured on an episode of the TV program 'True South' (Note: True South is a TV show broadcast on ESPC and the SEC Network. Its Purpose: "To tell an honest story about the complicated, flawed, and beautiful place we

call home. To do that we sidestep nostalgia and myth and stereotype to introduce viewers to Southerners who struggle and succeed, dream and fail, and then try again."). Although Mike uses some social media to market his products, most of his business comes from word-of-mouth recommendations, referrals from customers and testimonials from locals and visitors to the St. Augustine area. Not only does he sell fresh-catch seafood, he also has gotten into providing prepared foods from the products he delivers to walk-in customers (e.g., Minorcan Clam Chowder, Conch Chowder, Mullet and Salmon Smoked on site, Key Lime Pie, homemade cocktail and tarter sauces, platters and party trays to order.). The variety of seafood offered competes with any large commercial seafood dealer in northeast Florida, at competitive prices. He knows his Value, he knows his Customers, he knows his competitors and he knows his marketplace. This is a low key, no hype environment focused on selling only the best seafood (Note: You can even pick out a fresh fish, caught that day, and have them custom prepare a whole fish or filet.). He pays his very knowledgeable employees a fair wage for what they do (Note: Their loyalty and passion for the business is obvious as soon as you enter the store.).

- Crème' de la Cocoa: Although only ten years old, Crème' de la Cocoa already displays the attributes of sustainability. For the past few years they've been a winner and runner-up of the Food Network's many contests including winning the recent "Sweet Empire Winter Wars Team" contest and being a regular finalist with the teams competing in the Food Network's "Halloween Wars." Nils (Chef and Co-Owner) and Baily (Director of Operation and Co-Owner) Rowland have created one of the best High Value, High Worth Customer Experiences in St. Augustine Florida. Their Stated Mission is "to craft North Florida's best desserts while building meaningful relationships and unforgettable experiences" (Note: I personally testify that they meet their mission and exceed customer expectations.). Their fame is due to the quality of their products and the services they offer in conjunction with their unique creations and proven results. Their prices are at a premium, but well worth what they ask. They offer the best desserts I have had from across the entire world.
- I detailed the organizations above to provide a one stop location, that I am familiar with, that offers several examples of Affordability and Affordability Sustainment. I purposefully recommend that anyone who pursues Affordability and/or Affordability Sustainment should visit establishments and enterprises that demonstrate Affordability and Affordability Sustainment (Note: If you're interested, you may contact me and come visit St Augustine. I am more than willing to entertain and expose you to local Affordability organizations. Keep in mind, you may be able to "write it off" as a research trip seeking examples of continuous improvement and sustainability.).

St. Augustine's sustainability has been the result and outcome of robust flexibility with changing times, and responsive adaptability to culture, technology and the economy. As for Affordability, the Value provided by STA and the businesses I mentioned, meet and exceed the requirements of the targeted Customers that live and visit the city. STA Customers (i.e., Citizens and Visitors) can and are willing to pay the price set for the products and services offered, and although the cost of some of the products and services may be at a premium, they are all worth the price being paid as testified by the demand and success of the offerings. As revenues continue to exceed expenses, and the AIM of these establishments focus on Value and Customer success, profitability and prosperity will succeed and growth will occur and the future is bright.

SUSTAINING AFFORDABILITY

There are two dynamics at play for sustaining Affordability. One is the "Success Formula" (i.e., Purpose + Leadership + People + Resources and Tools + Design and Plan) and the other is the "Affordability Factors" (i.e., The Goal(s), Leadership, Strategy, Enterprise and Systems, Design for Affordability, Targeting and Aiming, Planning and Implementation, People, Process, Performance, Improvement, Sustainment). You may notice that Leadership and People both appear in the Success Formula and the list of Affordability Factors. From my experience, observation and research, Leadership and People are the two most important variables for Affordability. Consistency, predictability, respect, trust, integrity and dignity are all common core values of Affordability Sustainment.

It goes without saying (although I'm still going to say it), "Value, Customer and Cost with an AIM" are necessary to create and sustain Affordability. Products, services, systems and processes under the framework and foundation of Affordability set the groundwork for sustainment. Operations and the continuous improvement of the Enterprise play a critical role in establishing and sustaining Affordability.

The first step is to establish the environment for Affordability. Initially understand the Customers' requirements, wants, needs wishes and desires. Then define the Value (i.e., products and/or services) that meets and exceeds the Customers' demands (Note: If you have a product or service that already represents the Value you intend to deliver to the Customer, use a HOQ to evaluate and validate that Value offering.). Then determine how to create, produce and deliver that Value through a Value Stream. Then identify all goals and objectives. Then architect the systems and processes for the Enterprise and the Value Stream. Then target, aim, plan and implement (Note: Before going live you should experiment, test, pilot, validate and verify).

After being in operation for some period of time (i.e., at least three years), and in a profitable and prosperous business that measures performance, incorporates improvement and manages necessary changes, the time for instituting sustainment is at hand. Your "Affordability Venture" should be stable and predictable. Pursue standardization of all actions and activities associated with the business (Note: This includes all variables of the Success Formula and all components of the Affordability Factors.). Perform adjustments and modifications that make sense and align with your strategy. Keep in mind that success does not dictate to do the same thing over and over again without incorporating some change and improvement.

In a similar fashion as in the examples I presented earlier demonstrated, when Affordability is established and momentum is in motion, the practice of sustainment becomes normal operational mode. As part of the pursuit of continuous improvement and organization transformation, sustainment efforts should be aligned with strategy, product, service, systems, process and organization endeavors. Sustainment insures Value Delivery, Customer Loyalty, Profitability and Prosperity continue.

YOUR PATHWAY AND YOUR JOURNEY

If you're reading this book, and just beginning your Affordability journey, sustainment may seem like a small image in the vision of your distant future. However, if you're successful, after a few years, sustainment should be a part of your ongoing strategy. A three-year point is typically the time to decide on the ongoing viability of any business.

If you've been successful, looking for "tips and tricks" for the consideration of continued survival, take some time studying the Success Formula and the Affordability Factors. Set your strategy and course for the next three to five years to pursue and adopt these proven practices. If you haven't designed and planned a strategy, now is a good time.

If you've been in existence for some time, and you are looking for new or additional techniques to embrace, use the Success Formula and Affordability Factors to identify other organizations that demonstrate sustainability. Take time and visit these organizations. In fact, set up time to meet with the leadership to discuss the means and methods they use that keep them in business (Note: I have found that most organizations that have a long history are more than willing to share their story and keys to success and longevity.).

14 Where to Start

The secret to getting ahead is starting.

– Mark Twain

Believe in yourself. You are braver than you think, more talented than you know and capable of more than you imagine.

– Roy T. Bennett

STARTING AN AFFORDABILITY INITIATIVE

When I'm working with groups or individuals on transformation and change of organizations, the question I'm asked most frequently is, "Where do I start?" Often followed by, "How do I start?" With a final query, "What do I do now?" The sequence of where, how, what and when can be answered by: 1. Thoroughly Understand the Current State, 2. Research the Facts and Figures of Existing Condition, 3. Create short-term Solutions to try using available Resources, Tools, Methods and Models. Above all, establish momentum with quick wins using the "low hanging fruit" approach that will yield small energizing successes. And, at the same time, after establishing a vision and direction, work in parallel to achieve long-term victories.

A PERSONAL JOURNEY USING AFFORDABILITY AS A FRAMEWORK

For this section, I'm using a personal example that follows the common pattern observable in successful Affordability efforts. After living in the Atlanta, Georgia area for almost 30 years, we moved to Anastasia Island, St. Augustine Florida in the fall of 2018. As soon as we landed, I decided to employ a specific exercise routine for my aerobic, cardio and lower body major muscle groups (i.e., lower abdomen, hips, legs). My workout was comprised of a brisk walk on the beach for an hour each day. Starting in November 2018, through 2019 and 2020, and into 2021, my routine permitted me to exercise outside for around 300 days per year with only some interruptions caused by environmental conditions and personal commitments that prohibited me from enjoying my one-hour beach walk. Sometime in the spring of 2021, I began to be impacted by a sore right knee and a stiff left ankle. Each time the pain and stiffness appeared, I would have to take a day or two off to recover. Through the summer, my condition worsened and I began missing more days recovering, than exercising. In late September, I visited my Orthopedic Surgeon for diagnosis and treatment. After x-rays and a thorough examination, he presented me with three choices; 1. A knee operation, 2. Pain injections when required, 3. A low impact exercise routine.

DOI: 10.1201/9781003772934-14

For option #3, his thoughts were centered around employing a stationary bike, while my thinking focused on the idea of beach biking. I chose option #3. So, I had to consider, where do I start, how do I start, and what do I do now? The Affordability Factors were in the forefront of my thinking; Value (what was the value?), Customer (clearly, I am the customer), Cost (what will I have to pay and what will the overall expense be?).

During the month of October, and through most of November, I researched everything I could related to beach biking. First and foremost, I had to determine the best bike required. I discovered that there are at least 20 different versions and configurations of bikes being sold in the marketplace including: on road bikes, off road bikes, trail bikes, mountain bikes, racing bikes, cruisers, recumbents, fat tires, and even electric versions called "E-Bikes." Considering I was using the bike for exercise on the beach; I was able to narrow my choices down to 2" tire cruisers and 4" fat tires. Although both would work, I tried the cruiser but settled on the fat tire because of the type of sand we have on our beach. The fat tire that met my needs was priced at $700.00 (FYI: driving from my home to the beach where I parked to walk was at a cost of about $3.00 round trip and it took around ten minutes each way. For around 300 days a year, at $3.00 per trip, my gas expense was $900.00 so the $700.00 for the Fat Tire Bike was quite Affordable.). I went to the local bike shop near the beach and made my purchase the week after Thanksgiving.

I spent most of December riding various routes and recording the variables that affected each ride (i.e., time, temperature, sunny/cloudy, clear/rainy, wind speed, wind direction, tide level, sand conditions, low-average-high bike speeds, route). This was the research and understanding phase piloting, setting up and validating the system and processes. Toward the end of December, I was able to sketch a spaghetti diagram that illustrated a variety of possible routes with a regular target route. Knowing this diagram, I was able to use alternate routes when the wind was coming from different directions at different speeds, or when the tide level was high when the sand was dry, "sugary" and soft. Here is the original potential route diagram (Figure 14.1).

The solid line represents a common path for consideration for when the system and processes are stable and standard. The dashed lines show various other routes and options to consider depending on the wind, tide and other environmental conditions (Figure 14.2). The process of riding a bike is simple, but the route to choose was complex given the different variables that affected the ride. Toward the end of the month, after several days of trial and experimentation rides, I was able to render current state and future state value stream maps for a systems level view. Here is what the current state value stream map looked like: "Walk on the Beach."

For a 60-minute exercise session (value added time), it consumed 110 minutes or 1 hr 50 min of total time ("LT" or Lead Time). The exercise goal was to achieve an hour of exercise per day. The goal of "every day" was not met due to inclement weather and other personal time scheduling factors. For the prior few years, I had an average of 300 days of exercise per year as a result of about 65 days of schedule interruptions (i.e., rain, lightening, days below 50 degrees, wind more than 20 MPH,

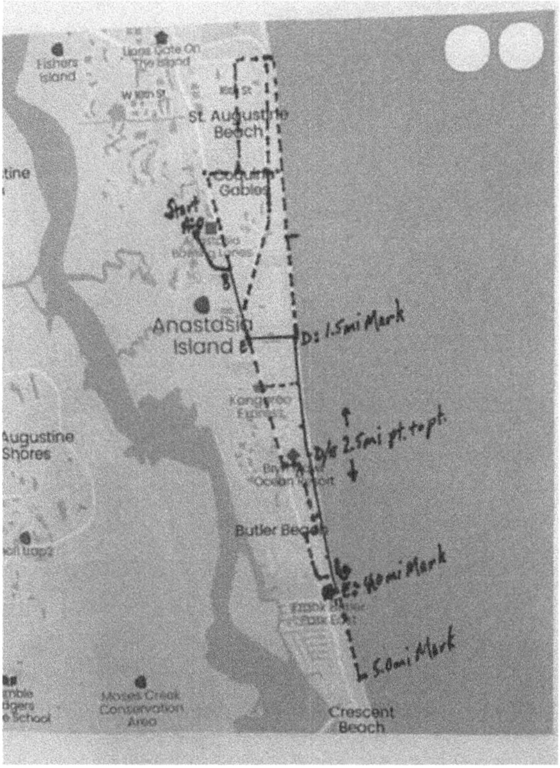

FIGURE 14.1 A spaghetti diagram of primary and secondary routes.

personal activities). The cost was negligible at a few dollars daily and the benefit of 300 calories burned each exercise event (Figure 14.3).

With this knowledge and the research data, I constructed a future state map for my first year of beach biking in 2022.

The goal of 60 minutes of exercise for 300 days per year remained the same, and the distance goal of 2,500 miles was set for 2022. Using the Lean Takt Time formula, I calculated the goal for speed (Time/Demand = 300 days × 60 min/2,500 mi → 18,000 min/2,500 mi = 7.2 min/mi). At 7.2 min per mile, or 8.3 MPH I was in range to accomplish all the goals established (Note: During December, I averaged 8–9 MPH in my trial runs.). It's also worthy to note that although the cycle time of exercise remained the same, the total time, or lead time for each exercise event was shorter, and the quality of each ride yielded more calories burned at a lower cost (i.e., LT = 96 min vs 110 min, Calories per ride range 440–770 vs 300, cost $800 bike vs ~$1,000 expense). Once again: faster, better, more affordable. In summary, I was able to gain personal time, burn more calories, lower my yearly exercise expense by biking rather than walking. More than anything, my knee was no longer painful (Note: To this day, my knee has not given me pain.), and my quality of life has improved.

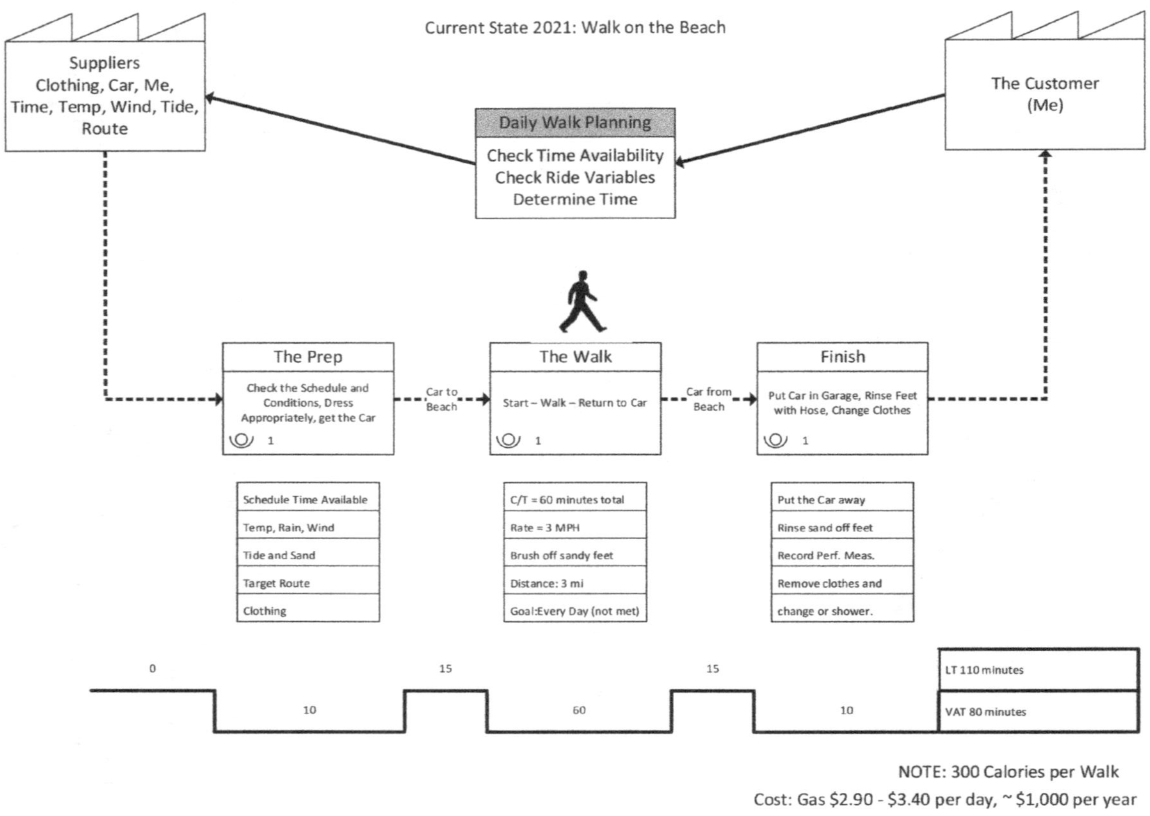

FIGURE 14.2 A current state value stream map of walking on the beach for exercise.

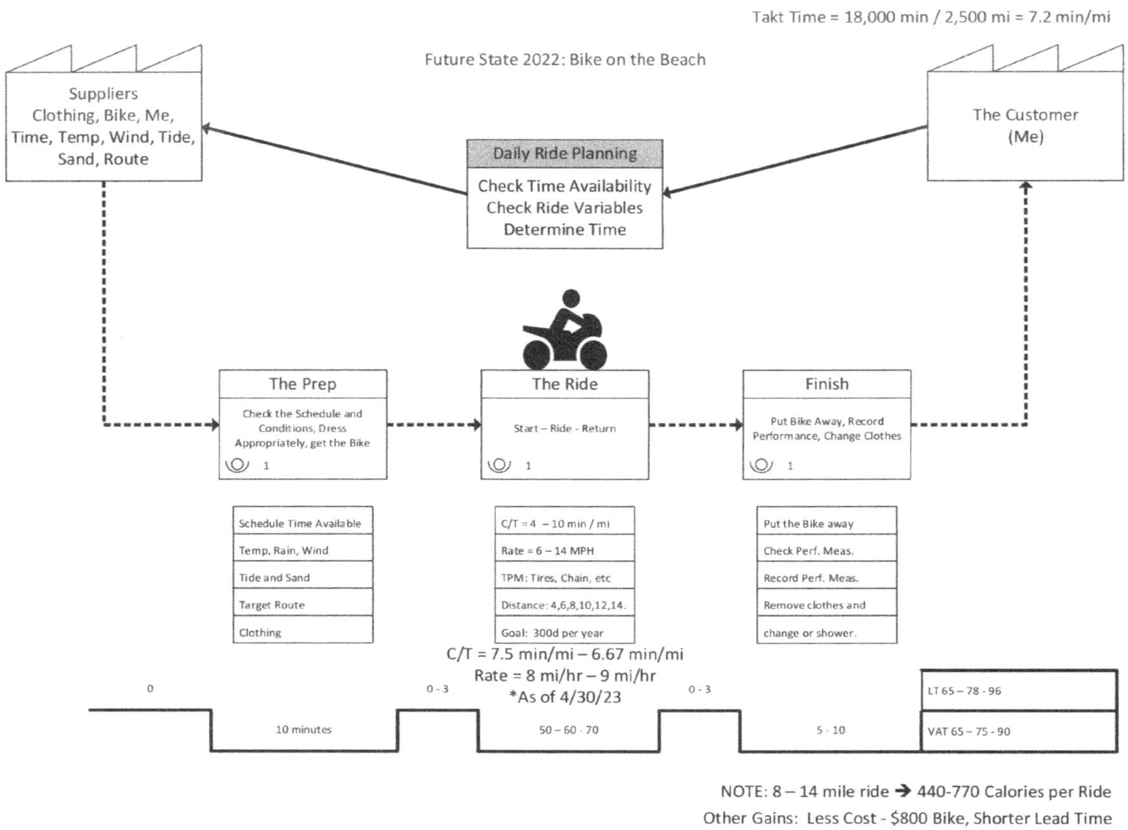

FIGURE 14.3 A future state value stream map of biking on the beach for exercise.

As far as detail, beginning on the very first day, January 1, 2022, I began recording every variable I encountered during each ride in an Excel Spreadsheet. After six months (knowing every factor for every condition), I standardized on two variables: time and distance. By the end of 2022 I achieved the following results: Total Miles—2,566 (Goal 2,500), Average Speed—8.4 MPH (Goal 8.3 MPH), Days—306 (Goal 300). For you statistical advocates out there, my Mean Daily Speed was 8.37 MPH with a Minimum Actual Distance of 4 mi on one day, a Maximum Actual Distance of 14 mi on a couple of days, a Std Deviation of 1.18 mi, creating an Upper Limit of 11.9 mi and a Lower Limit of 4.8 mi (Note: My average daily time for exercise was ~60 min). Using tools of Affordability, I was able to determine a process improvement, establish a direction with goals, align my resources for success, focus my aim, accomplish my targets, while going faster, improving quality of life and lowering my cost. At the end of my first year, the system was proven, and the process was stable and predictable, and the goals and objectives were met. I realized my initial calculations were on target (i.e., A Takt Time of 7.2 min/mi or a rate of 8.3 MPH allowed me to exceed the goal of 2,500 mi running at an average of 8.39 MPH).

The Table 14.1 summarizes my 2022 bike rides.

During the beginning of 2023, I isolated my annual goal to measure only a solitary metric of miles. I kept my Goal for Total Miles at 2,500 miles. I did continue to monitor my time per session, always seeking to exercise an average of 60 minutes per ride. I also kept the same yearly goal of 300 days of exercise. During January and February, I realized I was able to go faster and farther (i.e., Jan—28 days @ avg of 8.7 MPH for 243 mi and Feb—28 days @ 8.7 MPH for 243 mi as well) during each ride. It was clear to me, after the first two months, I had gotten stronger and more capable of going faster (i.e., 8.4 MPH average, up to 8.7 MPH average speed.). The trend at that time was such that by the end of the year, I would easily supersede the goal of 2,500 miles. Sure enough, by the end of the year, I had traveled 2,765 miles, on only 284 days of exercise. Although I did not meet the 300-day objective for 2023, I did greatly surpass the 2,500 mi

TABLE 14.1
Results of Year 1

Month	2022 Total	Daily Average	Days per Month
JAN-JUN	1268	8.2	151
JUL	180	8.2	22
AUG	251	8.7	29
SEP	237	8.8	27
OCT	227	8.1	28
NOV	200	8.3	24
DEC	203	8.1	25
TOTAL	2,566	8.39 MPH	306
	2,500	Goal	Total Days

target. The statistics told the rest of the story: Mean—8.89 mi, Lower Limit 4.31 mi, Upper Limit 13.47, Lowest Actual 3 mi, Highest Actual 15 mi, Average Time ~60 min.

For 2024, it was time to increase the goal and eventually render another future state value stream map for 2025 and beyond. I established a conservative goal of 2,750 miles, expecting to average just over 9.0 MPH per ride (Note: 9.0 MPH for 60 min X 300 days = 2,700 miles), possibly closing in on 10 MPH per ride by the end of the year to achieve at least 2,750 miles. After the first six months, I was beginning to average better than 10 miles per hour and often I was traveling more than 10 miles each trip (Note: Quite a few rides I felt very good and rode 11 mi–17 mi, using up more than an hour to do so.). By the end of 2024, I blew away the goal and accumulated 3,207 mi while averaging 10.02 MPH in 302 Days (Note: 60 minutes per session was no longer my average time!). As for the Statistics: Mean—10.02 mi, Upper Limit—13.60 mi, Lower Limit—6.45 mi, Actual Low—5 mi, Actual High—17 (Note: Most Days ~ 60 minutes of exercise). As the year ended, my final tally was 3,207 miles in 301 days.

About the middle of 2024, it became clear to me, my future state value stream map of 2022, was not only achieved, but surpassed and fully accomplished, in need of retirement as a completed current state value stream map requiring a new future state with new goals and objectives. After those first six months of 2024, my daily average distance exceeded 10 miles per day at a speed of around 10 MPH. As part of a natural pursuit of Affordability, I rendered a new future state value stream map. A new C/T (Cycle Time) with a new target rate according to a new Takt Time calculated using an increase in Demand (i.e., 3,000 mi, the new yearly goal). This VSM also allows for additional time beyond and hour for a range of 4 to 14 miles per ride (Figure 14.4).

Now, with new goals and objectives, starting with a conservative reality (Note: 6.60 min per mile or ~ 9.0 MPH for at least an hour.), I targeted a true 10 MPH average for 300 days or 3,000 miles per year. Since my strength and capability increased to a level much greater than 2022, I expected to bike at least 60 minutes with the option to add more miles depending on the environmental conditions and the energy I had during the ride. I no longer worried about time, since I was traveling 60 minutes or more, and I focused solely on distance (i.e., I had an expected minimum of 10 mi with an option of traveling further on any given day.).

2025 panned out as expected and projected: 3202 miles (Goal 3,000 mi), Mean—10.01 MPH, Lower Limit—5.83 mi, Upper Limit—14.31 mi, Actual Low—5 mi, Actual High—20 mi (NOTE: The Average Distance was a little more than 11 miles per trip.). The overall realized Value and unplanned gains included 176,110 calories burned that year, a stronger lower body major muscle group (i.e., hips, lower abdomen, legs), increased cardiovascular capability and aerobic capacity, with a psychological boost through release of endorphins, a greater quality of life and no more sore knees and ankles! My 2023–2025 results are as follows (Table 14.2).

I'm convinced that the change in my approach to exercise, and the transformation of my workout, has had a significant impact on my lower body, heart, soul and psyche. For example, today I biked 10 miles in 55:37, which is ~ 5.56 min per mile or

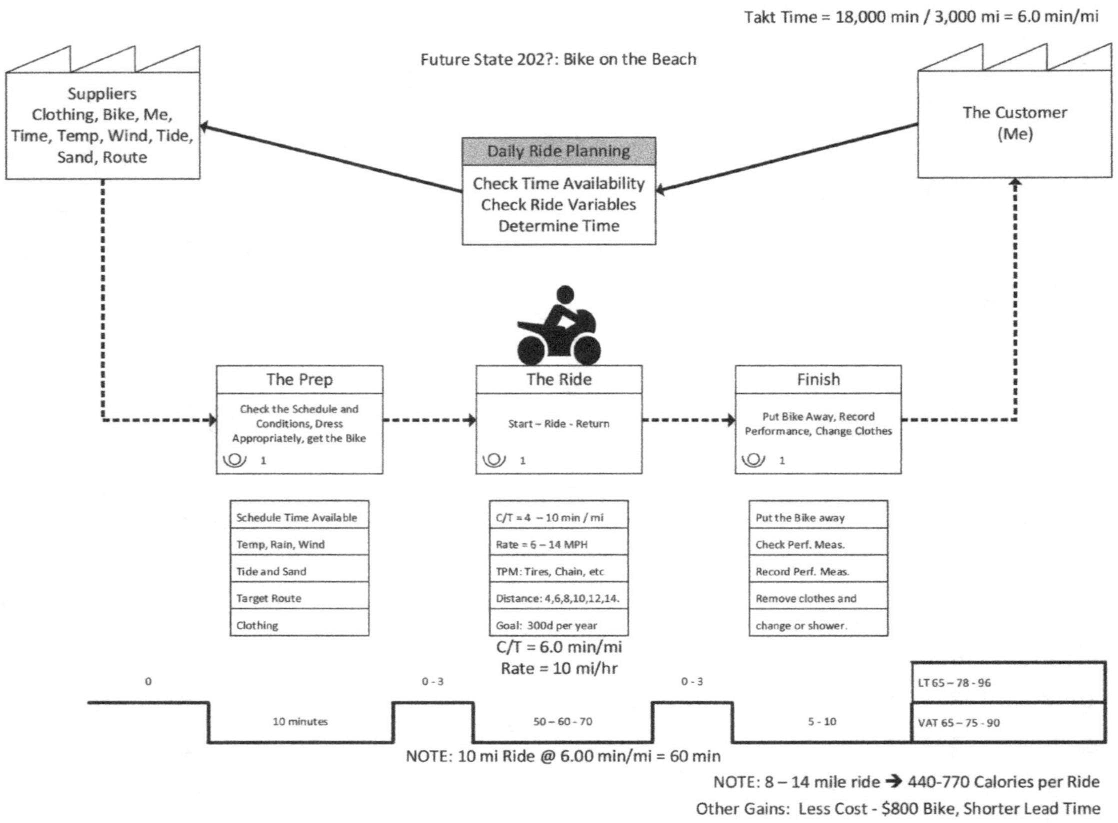

FIGURE 14.4 The next future state value stream map of beach biking for exercise.

TABLE 14.2
A 3-Year Record of Results

Month	2023 Total	Daily Avg.	Days per Month	2024 Total	Daily Avg.	Days per Month	2025 Total	Daily Avg.	Days per Month
JAN	243	8.7	28	253	9.4	27	160	10.7	15
FEB	243	8.7	28	252	9.7	26	268	9.9	27
MAR	243	8.4	29	308	10.3	30	276	10.2	27
APR	213	8.9	24	295	10.2	29	242	9.7	25
MAY	215	8.6	25	306	10.2	30	270	9.6	28
JUN	272	9.4	29	293	10.5	28	289	10.0	29
JUL	243	9.0	27	282	10.1	28	265	9.8	27
AUG	90	8.2	11	265	9.8	27	307	10.2	30
SEP	226	9.0	25	283	10.1	28	277	9.9	28
OCT	277	9.2	30	274	10.1	27	275	10.2	27
NOV	250	8.9	28	212	10.1	21	271	10.4	26
DEC	250	9.3	27	184	9.7	19	302	10.4	29
TOTAL	2,765	~ 9 MPH	284	3,207	~ 10 MPH	301	3,202	10 +MPH	289
	2,500	Goal	Total Days	2,750	Goal	Total Days	3,000	Goal	Total Days

a speed of 10.8 MPH (Note: With an 18 MPH wind coming from the north permitted me to go more than 12 MPH southbound with the wind and a little more than 8 MPH going northbound against the wind for an average of 10.8 MPH.). When you consider the enterprise view of my Value Stream Map, the current system is yielding the Value of health and exercise for the Customer (me) at a cost much less than the original Value Stream (i.e., using a walking process of more time and gasoline expense at @ $1,000.00 vs a lesser lead time and a one-time bike cost of $800 the first year with yearly maintenance of about $200 for the biking process.). Remember the mantra: Faster, Better, more Affordable!

I use this example as proof that Affordability primarily designed to address enterprises, can definitely be utilized by individuals for personal improvement to achieve a better quality of life. In addition to exercise, I have also used Affordability methods for remodeling my home to increase effectiveness in the kitchen and throughout the household, in the garage for workplace organization, and outside in my yard for curb appeal. Although it seems I'm obsessed with Affordability, I'm actually resolute in my conviction that this "science" and methodology can benefit anyone at work as much as it can benefit everyone at home. If you were to break down every example I provided you would likely recognize a pattern for creating solutions to complex problems is stable, consistent, predictable and sustainable.

SEEK SOLUTIONS

So, throughout this book, there appears a template for creating and implementing Successful Solutions. To begin, do your research and understand, then identify options to pursue with resources you might use, next design and plan for a solution that you can test, pilot and improve until you have a valid solution to implement, standardize and sustain. Other methods for solving problems do exist that follow a similar pattern; The Scientific Method, PDCA, A3, DMAIC, The Problem-Solving Model, and others. The method below is closely related to the classic 7 Step Method and time and time again has produced effective solutions for implementation:

1. Understand the Problem or Opportunity, Research, Find the root of what's happening.
2. Identify Options for a Solution, Identify Resources required.
3. Develop a Design and Plan for "The Solution" from the best option identified.
4. Experiment, Test and Pilot "The Solution" (go back to step 3 if "The Solution" fails).
5. Examine, Validate and Verify the Solution (go back to step 4 until the solution is proven)
6. Implement the Solution when its Proven and Complete
7. Standardize and Sustain the Solution, Reflect and Archive Lessons Learned

… continue looking for more opportunities. Take your opportunities back to Step 1.

Whether you're seeking a personal solution for a particular issue, a specific solution to a commercial problem, or even a comprehensive answer to a government crisis, a standard problem-solving process should be identified, implemented and utilized. Too often, a group wastes time deciding how to go about solving a problem, and not enough time solving the actual problem at hand. Be sure and use a standard problem-solving approach and process. Please be sure to use the correct people and resources that align best with the problem at hand. For most complex problems I recommend using teams of size 5 to 8 comprised of capable and competent individuals who have the knowledge, skills and experience in the problem area (Note: Teams larger than 8 in size are often more difficult when it comes to reaching consensus and compromise).

A CHECKLIST FOR YOUR READINESS BEFORE YOU START

Here is a 15-point checklist to evaluate your readiness for implementing Affordability.

- The Value you Deliver
- Purpose and Sense of Urgency
- Principles and Values, Vision and Mission
- Value Stream for Creating and Delivering Value: Product(s) and/or Service(s)
- Customer, Competition, Market
- Leadership and Management
- People
- Suppliers and Partners
- Enterprise, Systems, Processes, Procedures
- Strategy, Strategic Plan (Value Description, Customer, Market, Competition, UVP)

- Structure with Strategic, Operational and Tactical Alignment
- Tools and Resources
- Solution Design and Plan
- A Standard Problem-Solving Process
- A Scorecard for Performance Excellence

START WITH THE END IN MIND

Steven Covey taught,

> To begin with the end in mind means to start with a clear understanding of your destination. It means to know where you're going so that you better understand where you are now and so that the steps you take are always in the right direction.

Having such a perspective combines the current state of now with the future state of what where and how.

When I started my business, I began with some advice from mentors and some powerful philosophical thought from Dr. W. Edwards Deming towards the end of his life (1990–1993). Although I had only been in his presence a couple of times, I was able to mine a few key nuggets from his wisdom that I was able to apply to my journey,

- If you provide true Value, there's no need for marketing or advertising.
- If you provide Value that meets the Customers' needs, there's no need to 'sell it'.
- If you provide a price that the Customer is willing to pay, you will generate revenue.
- If you keep your overhead low, your profits will be higher.
- If you unleash people to solve problems, you will be successful.

Now that you might know where you're going, and you might have a sense of how to get there, and what you might do, I have some recommendations. I've gathered some personal recommendations for you to consider during your Affordability journey,

- Begin With the End in Mind
- Seek Guidance
- Stay Within Yourself, Seek Self Awareness
- Know What You Know, Seek Knowledge
- Research, Pursue Understanding, Learn
- Solve Complex Problems, Be of Value
- Try, Make Mistakes, Celebrate Successes
- Rally Strong Support, Leverage Teamwork
- Be Helpful
- Be Humble
- Be Thankful
- Be Perseverant
- Believe

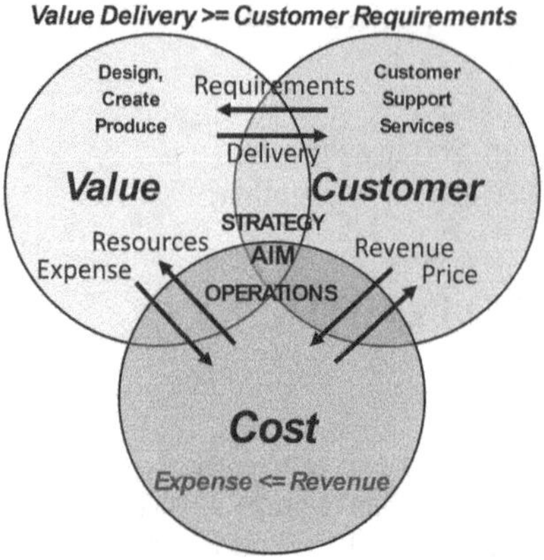

FIGURE 14.5 The dynamics of affordability.

Seek to understand, gather your resources, design a solution, build the solution, test and validate your solution ('until ready'), package and prepare the implementation, implement, maintain and sustain your solution. Continually seek to improve your Value and Value Stream.

YOUR PATHWAY AND YOUR JOURNEY

Your Pathway and Your Journey is unique to you and you alone. Affordability 2.0 is just one framework and foundation to consider for architecting and designing your way forward. It's clearly not "The Only Way," or "The Right Way," or even "The Way," but it is "A Way" that has proven to be successful for the many organizations I used for examples, standards and paradigms that accomplished their goals.

After studying and researching the factors that have made entrepreneurs successful, there has been one common thread: Delivering something of Value that meets and exceeds Customer requirements and demand at a price the Customer is willing to pay, and operating at an expense level that is less than the revenue generated by the income and revenue provided through Customer purchases.

I've heard Dr. Deming say, "There's no silver bullet, no magic wand, no instant pudding, and definitely no free lunch." If you follow the Affordability pathway, the journey will be difficult, and I promise there will be problems. Perseverance, persistence, flexibility and adaptability will be required. Use the means, methods, tools and techniques of Affordability (Figure 14.5).

Hopefully, you've gleaned and mined some 'nuggets' for use all along your journey. I wish you the best of luck and "*Godspeed*"!

Index

Note: Page numbers in italic denote figures and page numbers in bold denote tables.

www.ingramcontent.com/pod-product-compliance
Lightning Source LLC
LaVergne TN
LVHW011617161026
850215LV00029B/340
* 9 7 8 1 0 4 1 2 9 9 9 1 2 *